The Transputer Handbook

The Transputer Handbook

Ian Graham and Tim King

Prentice Hall

New York London Toronto Sydney Tokyo Singapore

First published 1990 by
Prentice Hall International (UK) Ltd
66 Wood Lane End, Hemel Hempstead
Hertfordshire HP2 4RG
A division of
Simon & Schuster International Group

Printed and bound in Great Britain at the
University Press, Cambridge

Library of Congress Cataloging-in-Publication Data

Graham, Ian, 1947–
The transputer handbook / Ian Graham and Tim King.
p. cm.
ISBN 0–13–929134–2 : $33.95
1. Transputers. I. King,Tim, 1954–. II. Title.
TK7895.T73G73 1990
004′.35—dc20 90-7072
CIP

British Library Cataloguing in Publication Data

Graham, Ian, *1947–*
The transputer handbook.
1. Computer systems. Parallel-processor systems
I. Title II. King, Tim
004.35

ISBN 0-13-929134-2

1 2 3 4 5 94 93 92 91 90

Contents

Preface

In this book we describe the software and hardware implementation of transputer parallel processing systems. We hope to bring together information from a multitude of sources in a more readily accessible form.

This book is not intended as a substitute for a study of engineering data when detailed design decisions are to be made, but should act as a guide to the capabilities of the transputer family and transputer-based systems. This should help the reader to choose the software and hardware solution that will best suit their problem.

Neither of us is an employee of INMOS Ltd, but we are both transputer users by choice, one in an academic environment, the other as head of a research-oriented software development company. Thus our opinions are our own, and do not always coincide with those of INMOS, and at times may be directly opposed. Despite this we would like to acknowledge the help provided by INMOS in the preparation of this book.

We would like to thank the following for permission to reproduce diagrams in the book: INMOS for Figures 2.1, 7.1–7.10, 8.1, 8.7–8.10 and 9.6; Meiko for Figure 8.11; and Parsytec GmbH for Figures 8.6 and 8.13–8.15.

The majority of our programming examples are in C, rather than occam, and we assume that the reader has a knowledge of C.

One of us (I.D.G.) thanks the University of Bath for a Visiting Fellowship, during the tenure of which this book was completed. We would also acknowledge the help provided by those transputer hardware and software manufacturers whose products are mentioned in the text.

Shepton Mallet — I.D.G.
1990 — T.J.K.

Notational Conventions

Internal processor registers and flags are shown in italics:

> *A, B, C, Error, HaltOnError*

Program examples and processor opcodes are shown in monospace typewriter font:

```
writeln("Hello World");
ldl  5
```

Signal and pin names are shown in the monospace font:

> `BootFromRom, notError`

Hexadecimal numbers are preceded by `0x`, and shown in the monospace font:

> `0xF` (4 bits)
> `0xF0` (8 bits)
> `0xF0F1` (16 bits)
> `0xF0F1F2F3` (32 bits)

Trademarks

UNIX is a trademark of AT&T.

VAX and PDP-11 are trademarks of Digital Equipment Corp.

IBM is a trademark of International Business Machines.

Computing Surface, In-Sun Computing Surface are trademarks of Meiko Ltd.

Helios is a trademark of Perihelion Software Ltd.

●, inmos, IMS and occam are trademarks of INMOS Limited. INMOS is a member of the SGS-THOMSON Microelectronics Group.

Chapter 1

Introduction

The rush to provide faster and faster computer systems seems to have taken us headlong from the classic IBM mainframe of the 1960s through the departmental minicomputers, typified by the VAX of the 1970s into the massive flood of personal computers during the 1980s. It is a well known but instructive cliché that the amount of computing now available in a wristwatch would have taken huge boxes of air-conditioned power thirty years ago.

The focus of the development of fast, innovative processors has shifted from the computer manufacturers to a small, and shrinking, group of semiconductor manufacturers. These semiconductor manufacturers continue to promise bigger (or rather smaller) and better devices each year, and there seems to be no limit to the amount of computing power that will eventually be available on your desktop.

Two questions need to be asked at this point: why do we need all this computing power, and how are we going to be provided with it?

1.1 Why more power?

This first question is often forgotten; computers just get faster and faster all the time. For many users, computers are already fast enough. Consider someone using a word processor or spreadsheet on a personal computer; so long as the text is displayed or the spreadsheet updated within a second or so then getting a faster computer so that the update takes half a second hardly seems worthwhile. These users have no real need of more computer power, and are happy with their system.

Other users can clearly see room for improvement. If the word processor has been replaced with a desktop publishing package then the speed with which the system responds may well become more critical. Unlike the previous system a character is represented on the screen by a pixel pattern defined by a chosen font. The computer system has to perform a significant amount of extra work in this case. All the word-processor software had to do to display a character was to load its value into a hardware register; the character generator built into the display circuitry would then cause the correct pixels to be sent to the screen.

The modern flexible font system places a much greater burden on the software. A software routine must locate the correct bitmap for the character, size and font required, possibly generating it from a stored set of Bezier curves. The computer must then set each individual pixel on the screen to represent the required character. All of this work is needed just to display a single character. There is then considerable extra effort required in order to scale pictures, move columns of text around a screen, change the width and format of columns and so on. The amount of extra work to be performed ranges from 100 to 1000 times more.

It is hardly surprising that a computer capable of providing excellent performance when faced with simple tasks suddenly shows its limitations when asked to handle a much more complicated program. What is sometimes not appreciated is exactly how much more complex desktop publishing is than simple word processing. So far as the user is concerned it is a simple step up from limited fonts and poor quality to multiple proportional fonts and high quality.

Current software users are becoming more critical of the user interface provided. Originally it was perfectly acceptable for a software package to have obscure commands that had to be typed before it could be persuaded to do anything. Users were not to be trusted with anything as complicated as a computer, unless they were willing to spend hours reading the manuals and attending training courses. It was also considered a matter of honor that individual companies developed a certain house style, so that pressing a particular key on one package might have a devastatingly different effect from the same key in a package from a rival manufacturer.

Modern developments have shown that good software should be intuitive. If the user has learned one program package then the next package, albeit from a different software house, should behave in a similar fashion. Standard screens have been replaced by windows, overlaid on other windows. More than one program can be run at the same time. The screen itself is capable of displaying complex graphics and variable-sized fonts. A mouse is used to select from iconic images of programs, while pull-down menus are used to select the different options.

This general improvement in the user interface has had yet another major impact on the amount of computer power required behind it. Graphical systems have become the standard for every computer user, rather than being used occasionally by specialists in order to plot a diagram or draw a graph. Graphics resolution has improved dramatically from 320 $\times$ 200 pixels on the screen to 1024 $\times$ 768 or more. All of this uses more and more computer power.

On the general scale, more computer power enables us to produce better models of the real world. These might be higher quality graphical models, quicker statistical models or more accurate engineering models. As our requirements for these models increase then our appetite for computer power also increases. Consider areas such as computer simulation, where a full simulation of an intricate piece of engineering such as a power station must be performed in order to verify the design. Even on the largest of present computers such simulations can take years.

There are many areas of potential computer applications that are not currently being addressed because the technology is simply too slow. Good examples here are concerned with getting the computer to do anything that the human brain is good at, such as voice recognition, image analysis and pattern recognition, artificial vision, and natural language understanding. We are a very long way from the science fiction of having a robot come into the room, recognize us and understand our natural speech. Current research into the software solutions to these problems suggests that huge amounts of computer power will be required.

1.2 How do we get more power?

The preceding section put forward some arguments as to why more computing power is required now, and why we see no saturation of this requirement in the future. This section describes some of the problems in making computers faster and faster.

The underlying and absolute physical limitation is the speed of light. Certain semiconductor manufacturers seem to be suggesting that their particular technology has a rosy and unencumbered future, with faster and totally compatible models coming out each year for ever. This is, of course, impossible. The technology wall presented by the speed of light is at the top of a technology hill. Not only is it impossible to scale the wall, it also gets extremely difficult even to get close to it.

Desktop personal computers used to be based on microprocessors with an 8 MHz clock speed, and recent developments have seen the introduction of 33 MHz chips. At first glance this can be seen as the gradual, but irresistible, onwards march of technology. A more detailed examination shows that even this change has led to more problems than might first have been assumed.

Consider the problem of attempting to design a memory interface. As anyone who has bought a personal computer will know, a microprocessor with a clock rate of 16 MHz will not run twice as fast as one with a clock speed of 8 MHz. A normal dynamic memory chip will have an access time of between 80 and 100 nsec, and the faster the chip the more expensive it gets. It is possible to use static RAM devices, which have a much shorter access time, but which are perhaps ten times as expensive, and are not available in such high density packages as dynamic memory. Thus the memory is unlikely to be able to keep up with the 16 MHz processor. A common solution is to provide a cache memory of fast RAM, but this is expensive, and relies on the locality of programs and data.

Bus structures become much less useful as processor speeds increase. Until recently it was common for a microprocessor system to consist of a CPU board and some memory boards connected by a high-speed bus. However, processors are now sufficiently fast that memory access is unacceptably slowed by the physical layout of the bus and its associated capacitances. Even high-speed peripheral interfaces can no longer reside on the bus.

Part of the approach to solving these problems has been to place more of the required circuitry on a single chip. On-chip devices, such as cache memory, memory management units and floating-point units, are preferable as no external buffers are required and the distances involved are very small. Complex single-chip devices have been made possible by the development of very large scale integration (VLSI), which in its turn is encouraged by the requirement for more and more complex devices. The recent Intel processor, the i860, may have a reduced instruction set architecture but with more than one million transistors it is hardly a simple device. These complex chips have their own problems. The chips are difficult to manufacture with high yields, and are therefore expensive. They dissipate a lot of heat, which has to be removed from the chip in some way, and they require fast memory and generate fast signals, which places strict limits on board design.

An analogy to this problem is that of trying to get a drink in a crowded bar. The bartender starts to get overwhelmed, and so a replacement with more experience is brought in. He works faster, but still cannot cope with the increasing demand, and so another replacement is found who has been training for the Olympics. Although she is now racing from one side of the bar to the other, she still cannot cope. Efficiency is lost, more drinks are spilled and glasses broken.

The solution, of course, is to replace the single bartender with five or six who share the work. If the amount of work increases yet again, more staff are brought in to continue the service.

In order to produce more and more powerful computers we must provide power through multiple processors, working together to achieve a solution to a large shared problem. The techniques of parallel computation are applicable to all scales of processor, from the smallest to the largest, and will not be made obsolete by new technology, as we will always require more power than can be provided by any single processor, whatever its technology.

1.3 Types of parallel computer

There are many ways in which to define a taxonomy of parallel computers. An early attempt by Flynn (1966, 1972) divided computer systems into four major categories based on the number of instruction and data streams that are processed simultaneously. Flynn's categories are as follows:

- Single instruction stream, single data stream machines (SISD).
- Single instruction stream, multiple data stream machines (SIMD).
- Multiple instruction stream, single data stream machines (MISD).
- Multiple instruction stream, multiple data stream machines (MIMD).

The SISD architecture still describes most of today's large and small computers, where a single program executes using a single set of data. SISD machines can potentially have multiple functional units, such as vector processors or pipelines, as is common in today's 'uniprocessor' supercomputers.

SIMD machines execute the same instruction stream simultaneously on many sets of data – examples are array processors such as the ICL Distributed Array Processor (Reddaway, 1984). Systems with this architecture are restricted in the problems that they can solve, but are particularly suited to operations on large matrices.

MISD machines potentially execute many instructions in multiple streams on a single data stream. However, no examples of this machine organization seem to exist.

We will be concerned here principally with the MIMD machines, where multiple sets of possibly different instructions are executed concurrently on multiple data sets. This is the most general model, as it is possible to make a MIMD machine behave as either a MISD or SIMD machine by suitable programming.

The Flynn taxonomy provides a crude separation of machine types. For a more sophisticated analysis and alternative taxonomies the reader is referred to Hockney and Jesshope (1988) and Krishnamurthy (1989).

1.4 MIMD architectures

In this book we discuss the hardware implementation and programming of machines of the MIMD type. Such systems are already common, and two main architectures have emerged which differ in how the processors communicate to share data and control. It is clear that the processors should communicate, either with each other or with a controlling processor, in order that the computing task may be distributed over the parallel system and the results of the computation collated.

1.4.1 Shared-memory machines

In shared-memory systems the processors use a common pool of memory, normally with some hardware constraints to avoid memory access conflicts. Tasks communicate through shared variables, and code and data may also be shared by tasks running on different processors. One clear disadvantage of the shared-memory architecture is that a limited memory bandwidth must be shared by all the processors in the system. Thus as each new processor is added, the system performance does not improve linearly, but follows the law of diminishing returns. The amount of degradation is clearly very dependent on the application, and on the details of the memory architecture. The memory contention problem thus limits the number of processors that can be added to the system, and the large memory bandwidth required in the main memory adds to its cost.

1.4.2 Distributed-memory machines

Alternatively, each processor may have its own local memory, and communicate with the others over hardware channels or links. The total memory bandwidth and instruction execution rate thus increase linearly as more processors are added to the system. The principal limitation of this message-passing architecture lies in the communication links. If in a system of n processors each must communicate directly with all of the others then $n(n-1)/2$ bidirectional links would be required. This would result in an impossibly complex architecture with even a modest number of processors. Thus practical solutions limit the number of hardware links to each processor, and provide either software or hardware message switching so that data and control information can be transferred between processors as the application requires. If this communications system and the topology of interconnection are not well designed, then communications overheads can severely limit the overall performance of the parallel machine.

1.5 The transputer

The INMOS transputer is the first single-chip microprocessor that provides a high-speed processor, fast inter-processor communications, and explicit support for multiple processes and multiple processor systems. The design aims were for a device that would be used in multiple processor message-passing systems, where each processor had its own physical memory, but with support for multiple shared-memory processes on each transputer. What the transputer does not provide, therefore, is any memory management on chip, or any support for off-chip memory management devices.

The intended application areas range from process control, with 1–50 loosely coupled processors, workstations with 4–16 processors, to computing accelerators and supercomputers with over 256 processors. The transputer is a reasonably recent development and yet it has already had a profound impact on the way in which certain types of problems are solved.

Chapter 2

The Transputer

This chapter describes the architecture of the transputer family. In fact there is a fairly wide range of transputer devices available, but the family resemblance is very strong, and they all follow the basic outline described here. The family members are described in summary at the end of this chapter; more detailed information on the properties of individual devices is given in later chapters.

The basic components of all microprocessors are the instruction execution and decoding unit, often called the central processing unit or CPU, and an external memory interface or EMI. The CPU reads instructions from memory via the EMI and executes them sequentially. These two components make up the complete internals of older chips such as the Z80 or the 68000. Many modern chip designs concentrate on making this standard processor and memory interface combination run as fast as possible. However, increasing sophistication in VLSI technology has allowed designers to increase the level of system integration, and to include on-chip functions that previously required external coprocessors. The most important additions are memory management, floating-point arithmetic, and instruction and data caches. In each case the provision of these functions on chip provides a substantial performance improvement over the same function implemented in external hardware.

Memory management hardware provides a mapping between the large virtual address space of the processor and the relatively small amount of physical memory. Each process can have a separate virtual address space and a different mapping into physical memory, and thus the memory used by one process can be protected from access by other processes and can appear to grow beyond the physical memory limits. The address translation hardware can be made to run much faster when the memory management module is integrated on the same chip as the CPU and memory interface.

The same is true of floating-point coprocessors. The performance of an external coprocessor may be limited by the speed at which the main processor can load it with instructions and operands and store the results in memory. The external coprocessor interface also tends to share the same signal lines as the EMI, and to

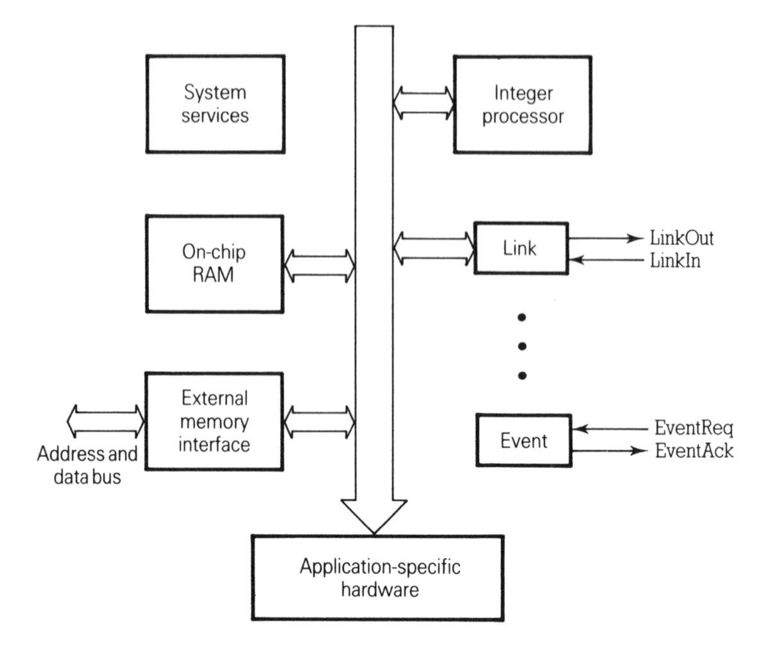

Figure 2.1 The generic transputer

reduce the memory bandwidth of the system. This has led the designers of some of the most recent high-performance microprocessors, such as the 80486 and i860, to include on-chip floating-point units.

Another common addition on chip is a cache of some kind, which allows frequently accessed data and code to be held on chip. Sometimes this cache is implemented for instructions only, as on the 68020; or with separate instruction and data caches as on the i860. The transputer family uses a rather different approach at present, having on-chip RAM that can be used in exactly the same manner as external memory, except that accesses to the internal memory are at least two to three times faster. Unlike a cache, this requires the programmer to make explicit use of the fast RAM, and to decide if it should contain code, or data, or a mixture of both.

The major extra function built in to every member of the transputer family is that of on-chip communications. It is this single item that makes the transputer unique, as every transputer comes equipped with the ability to communicate with other transputers. It is thus very easy to construct arrays of processors working together as a MIMD computer.

In summary, a transputer (Figure 2.1) is a single-chip VLSI device with processor, memory and communications links, which represents a slight deviation from current microprocessor technology. Along with various support devices it is manufactured by INMOS Ltd, part of the SGS-Thompson Microelectronics Group. The first transputer product, the IMS T414, was introduced in 1986, and by December 1989 there were ten processors in the family. The common features of all present transputers are as follows:

- High-speed integer processor with microcoded process scheduler.
- On-chip fast static memory.
- Up to four links for communication with other transputers.
- Internal timers.
- External memory interface.

However, the transputer lacks one feature that we have begun to expect in more conventional microprocessors: there is no support for memory management, or for virtual memory. This lack is a result of the design goals of the transputer. It is a device intended for the implementation of multiprocessor systems, where each processor has its own local memory and processors communicate by message-passing along fast links. Thus protection of the address space of one process from the actions of another can be provided by putting the two processes on separate transputers, rather than by a memory management scheme.

The transputer family consists of a number of different processors that have been optimized for different tasks. However, the overall architecture of the transputer makes it easy to construct and to program systems containing a mix of these transputer types. The transputer links make hardware interconnection fairly easy, and the transputer instruction set is designed so that programs can be written that are independent of the word length of the processor, even at the binary level. Transputers will normally be used in multiprocessor systems and thus the performance of the individual processor is not so critical. If more processor power is required, more processors can be added with very little modification of the software.

2.1 Microprocessor architectures

Current microprocessors fall into two main groups. On the one hand lie the complex instruction set computers (CISC) where a rich instruction set is available to manipulate data both in memory and in registers. The alternative approach is based on the reduced instruction set (RISC) philosophy. Here a small number of instructions are available, most acting only on data in registers. The logic behind the RISC approach is that it is possible to construct a faster computer by keeping it simple.

The transputer CPU does not fall readily into either category. It has a simple instruction set which means that it tends to be viewed as a RISC processor. It is, however, much more than a RISC processor because of the extra functionality built into the chip to support high-level concepts such as processes, timers and inter-process communication.

It is useful to review some of the more familiar microprocessor families at this stage in order to compare them with the transputer. The two major families of CISC processors started with the 8086 from Intel and the 68000 from Motorola. The Intel range has expanded to include the 80486, whereas Motorola have the 68040 at the top end of their range.

The Intel 80386 is a popular 32-bit chip providing integer-only operations. A floating-point coprocessor may be added externally. There is a set of registers of which many have special purposes in certain instructions. The chip provides a way of emulating the older 8086 processor with its segmented architecture, but in native mode a full 32-bit linear address space is available. There is a range of specific opcodes intended to speed up certain operations such as copying strings as well as the standard arithmetic and logical operations. Many operations may be performed directly to memory without loading the value into a register. The chip contains an integral memory management module.

The Motorola 68030 at first sight seems slightly more symmetric. There are sixteen general-purpose registers but these are divided into two types, called data and address registers, and many operations can only be performed in one type. Special addressing modes use the registers in different ways and can decrement or increment them as a side-effect of other instructions. The chip is integer only but an external floating-point coprocessor adds a separate set of floating-point registers. There are no special string instructions, instead an instruction cache is provided which ensures that any short loop becomes resident on chip. A data cache is also used to speed performance. The linear memory address space is handled by an integral memory management unit. Many standard operations, such as addition, can load one argument from memory but always leave the result in a register. The instruction set includes a number of special case operations to manipulate such things as program modules.

Both Intel and Motorola have recently announced their entry into the RISC camp in the shape of the Intel i860 and the Motorola 88000 microprocessors. The 88000 has thirty-two general-purpose 32-bit registers and two completely separate address buses for instructions and data. As with many RISC designs, all operations are handled in registers; two specific load and store operations are the only way of accessing memory. A floating-point unit is provided on the chip, whereas cache and memory management are handled externally by other members of the 88000 family.

Another RISC chip is the SPARC design from Sun Microsystems. This is a classic RISC design which has been used to extend a range of computers previously based on 68020 processors. The logic behind a RISC processor is that most programs are written in a high-level language and compilers are not very good at using special-purpose instructions. Most processors spend 90 percent of their time executing 10 percent of their instructions; the next step is to speed up those instructions and leave out the complicated ones, such as multiplication, which can be provided by software. The SPARC design provides a simple but fast processor based on three-address instructions so that each opcode is of the form 'add R1 to R2 and place the result in R3'. The result is that each instruction is 32 bits wide; the problem with the RISC approach is the processor can spend a large amount of its time reading these simple instructions in order to execute them.

The SPARC design has one interesting design feature not generally found on other commercial RISC chips: the concept of a 'register window'. A compiler can make good use of a small number of registers and ensure that few redundant stores and loads from memory are made. However, as soon as a procedure is called the current register set is dumped because the compiler does not know what registers are used in the procedure; when the procedure returns the registers are restored from memory. The SPARC architecture provides a different set of registers for the new procedure which are allocated from a circular buffer. Only when the entire register set has been used up does the compiler need to save the previous contents.

2.2 Transputer architecture

The internal design of the transputer is unlike that of any of the preceding processors. A central concept of the transputer is that of the process and this can be seen throughout the instruction set. A process represents an individual thread of control and the transputer switches between running processes to provide the illusion that they are all running simultaneously. This process switching is sometimes called multi-tasking and is normally handled by an operating system, but in the transputer it is implemented totally by the hardware and microcode of the processor.

The register model of the transputer is shown in Figure 2.2. All registers are either 16 or 32 bits long, depending on the word length of the transputer. Registers A, B and C form an evaluation stack and transputer instructions are designed around the use of this stack rather than using general-purpose registers. A stack depth of three provides a good compromise between the ability to evaluate most expressions on the stack and having as little as possible to save when a context switch occurs.

Register W is the workspace pointer, a pointer into the local variables associated with the currently executing process. Many instructions refer to data by their offset from the workspace pointer. The instruction pointer register I points to the next instruction to be executed, and this corresponds to the program counter PC in more conventional processors. The operand register O is used in the construction of instruction operands.

2.2.1 The process scheduler

A unique feature of transputer architecture is that it contains a microcoded scheduler which maintains two process queues, one at high and one at low priority. Processes in the high-priority queue are allowed to execute until they terminate, or

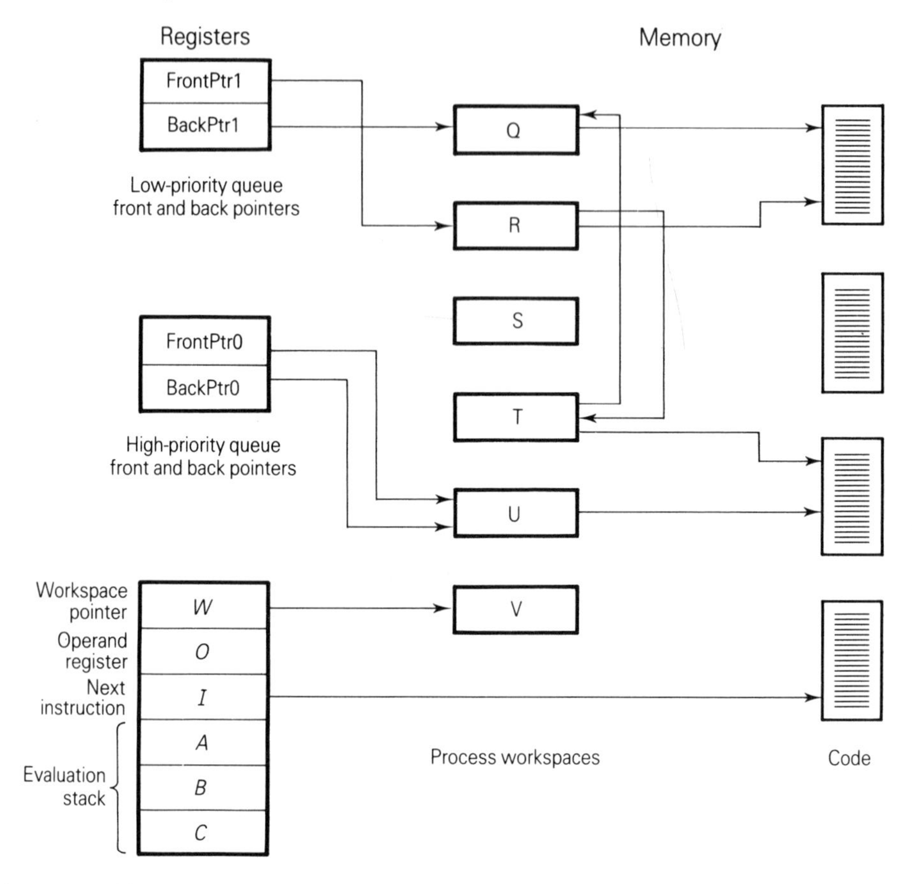

Figure 2.2 Transputer registers

require input or output. However, low-priority processes are automatically time-sliced at about 1 msec intervals, and can also be interrupted by a high-priority process. These queues are implemented as linked lists through the workspaces of the active processes, the front and back of the queues are pointed to by two pairs of registers, one for each priority. The mechanics of process switching are discussed in detail in Section 3.7.

Associated with each priority level is a timer and a timer process queue. The high-priority timer ticks once every microsecond, the low-priority timer every 64 μsec, or exactly 15,625 ticks per second. The number of bits in the timer registers depends on the processor model: for 16-bit processors the high-priority timer will cycle in about 65.5 msec, the low-priority timer in 4.2 sec; for 32-bit processors the high-priority timer cycles about every 4,295 sec (1.2 hours), the low-priority timer about every 76 hours. The timer registers can be read directly, or the scheduler can be instructed to queue a process for execution after the timer has reached a certain value. The timer queues are again implemented as linked lists within the workspace of the waiting processes, but the entries are sorted into time order. The front pointers to the timer queues are stored in reserved locations.

Another important feature of the scheduler is its ability to select one of a group of processes for execution, depending on the occurrence of some event. These events may be the completion of a data input, the expiration of a timer period or an external interrupt.

2.2.2 Communications

As the transputer has been designed for use in message-passing parallel computers it has strong support for inter-process communication. This has been designed so that there is almost no difference to the programmer whether the communication is between two processes executing on the same or on different transputers.

Data is transferred between two processes on the same transputer by copying data between their memory spaces. This transfer is synchronized by means of a channel control word, which is a single word anywhere in memory. Communication only takes place when both the inputting and outputting process are ready, and processes waiting for communication take up no processor time. Thus two transputer processes may synchronize by passing a message between them. Communication between two processes on different transputers uses just the same mechanism, synchronized by special channel control words that lie in reserved locations in low memory. The same instructions are used to set up the transfer, and the link engines in each transputer take care of the direct memory access (DMA) transfer between link and memory, with no processor intervention.

Each of the presently available transputers has either two or four external links. These are full duplex communications links that can exchange data with other transputers at 5, 10 or 20 million bits per second (Mbps), the speed being selected by voltages applied to three pins on the transputer package. The bidirectional data rate that can be achieved over the links depends on the types of processors exchanging data, and the link speed, but can be up to about 2.4 Mbytes per second for T800 transputers with 20 Mbps links.

The link data are transferred as a serial byte stream, each byte being acknowledged by the receiving transputer. No attempt is made to detect errors on the link; it is assumed that the communications medium is error-free, or that higher levels of communications will detect and correct errors.

2.2.3 Interrupts

The transputer has a single source of external interrupts, the `EventReq` input. The programming interface of this input is implemented so that it appears as another channel control word, located in low memory. A process instructed to wait on the event channel will be descheduled until the EventReq input is taken high, when it will be rescheduled. When this occurs the event handshake output EventAck is driven high by the processor. In order to provide a rapid response to interrupts the process waiting for event input must have high priority, and be the only high-priority process running. If these conditions are met the worst case interrupt response time is about fifty processor cycles. This fast interrupt response is obtained by making long instructions interruptable.

2.2.4 Memory

The present members of the transputer family have 2 or 4 Kbytes of on-chip fast static RAM, which occupies the very bottom of the memory space. Some of this is reserved for processor microcode functions. This is an inadequate amount of memory for the majority of applications, and thus each transputer has an external memory interface. Reads or writes to internal memory are fast, requiring only a single cycle of the processor clock. External memory access is much slower, requiring at least two processor cycles and usually four or five in dynamic memory systems.

Memory addresses are signed, thus the memory of a transputer starts at the lowest possible negative number, MinInt, which is 0x8000 on a 16-bit and 0x80000000 on a 32-bit transputer. It runs through zero to the largest possible positive number, MaxInt, 0x7FFF or 0x7FFFFFFF. Several locations at the bottom of memory are reserved for processor functions, as shown in Table 2.1. In this table the addresses are shown as word offsets from MinInt, the actual byte address depending on the word length of the transputer. The lowest eight words contain the channel control words for the four external links, with the event control word in the ninth location. The next two locations contain the front pointers for the high-priority and low-priority timer process queues. Above these are the seven words that store the processor state when a low-priority process is interrupted by a high-priority process, the only occasion when the processor state has to be saved.

2.2.5 System services

The system services interface includes signals necessary to reset and boot the transputer, to set the speed of its processor and its links, to signal errors and to respond to external events. The transputer can be reset into either of two modes, determined by the value of the Analyse signal when Reset is asserted. If reset with Analyse inactive the transputer will boot from memory if the signal BootFromRom is connected high. If BootFromRom is low the transputer will listen on its links, and can accept a bootstrap program or commands to write (poke) or read (peek) memory locations via the links. The Analyse input is used as a debugging aid. If Analyse is taken high while the transputer is running it will halt very shortly after, and the links will become inactive. Reset may then be asserted. When Reset

Table 2.1 Reserved memory locations

Word address	Name	Use
MinInt+28	MemStart	T805, T801, T800, T425, T225
...		
...		reserved for extended
...		instructions
...		(not T212, M212, T414)
...		
MinInt+18	MemStart	T414, T212, M212
MinInt+17	EregIntSaveLoc	
MinInt+16	STATUSIntSaveLoc	
MinInt+15	CregIntSaveLoc	
MinInt+14	BregIntSaveLoc	register save
MinInt+13	AregIntSaveLoc	area
MinInt+12	IptrIntSaveLoc	
MinInt+11	WdescIntSaveLoc	
MinInt+10	TptrLoc1	low-priority timer
MinInt+9	TPtrLoc0	high-priority timer
MinInt+8	Event	event channel
MinInt+7	Link3Input	
MinInt+6	Link2Input	
MinInt+5	Link1Input	
MinInt+4	Link0Input	link control
MinInt+3	Link3Output	words
MinInt+2	Link2Output	
MinInt+1	Link1Output	
MinInt	Link0Output	

comes low again the transputer will be in its reset state, but the registers will contain information on the state of the machine when it was halted by the assertion of `Analyse`.

2.3 The transputer family

The transputer family consists of three main groups: the 16-bit T2 series; the 32-bit integer-only T4 series; and the 32-bit T8 series, which have an on-chip 32/64-bit floating-point processor. Within these groups processors are distinguished by the amount of on-chip RAM, the number of links, details of the instruction set and the type of memory interface. However, all transputers conform to the general architecture model shown in Figure 2.1. Here we will introduce the present members of the transputer family, as summarized in Table 2.2; a more detailed description is given in Chapter 7.

Table 2.2 The transputer family (December 1989)

	16-bit transputers			
	T212	T222	T225	M212
Word length	16	16	16	16
Internal RAM	2K	4K	4K	2K
Number of links	4	4	4	2
Extended instructions	n	y	y	n
Debugging instructions	n	n	y	n

	32-bit transputers					
	T400	T414	T425	T800	T801	T805
Word length	32	32	32	32	32	32
Internal RAM	2K	2K	4K	4K	4K	4K
Number of links	2	4	4	4	4	4
Hardware FPU	n	n	n	y	y	y

2.3.1 Sixteen-bit transputers

The T212 was the original 16-bit transputer, with 2 Kbytes of internal static memory; the T222 is a more recent device with 4 Kbytes of internal memory and an extended instruction set. The T225 is almost identical to the T222, but contains extra instructions to support debugging. The M212 is a derivative of the T212 with a built-in disk interface, 1 Kbyte of RAM and 4 Kbytes of internal ROM containing disk controller firmware. The M212 has only two external links, the other two are used as part of the disk interface.

2.3.2 Thirty-two-bit transputers

The T414 is the original member of the transputer family. It has a 32-bit integer processor with 2 Kbytes of internal RAM. The T425 is an updated version of the T414, with block move instructions. This processor introduced instruction set extensions to support debugging, and is intended eventually to replace the T414. The recently announced T400 is a simplified and low-cost version of the T425, with only two links and 2 Kbytes of on-chip memory.

The first transputer specifically designed for numerical applications was the T800. This has 4 Kbytes of internal memory, and a floating-point unit (FPU) capable of operating to the IEEE-754 specification on 32- and 64-bit numbers. As the FPU is on chip the floating-point performance is excellent. The T805 is a development of the T800 with the same debugging support in its instruction set as the T225 and T425. The T801 is similar, but has a high-speed static RAM external memory interface, rather than the programmable interface of the other members of the T8 and T4 series. This makes it able to run faster, but requires more expensive and less compact memory devices. The 32-bit transputers are available in a range of clock speeds from 20 to 30 MHz.

2.3.3 Future transputers

INMOS have recently announced their plans for the development of a much higher speed transputer, code-named the H1. This device is intended to have a processor roughly ten times the speed of present transputers, and links capable of an aggregate data rate of 80 Mbytes per second. The processor will have support for standard operating systems, presumably some form of memory management, and a simplified interface to dynamic RAM. It should be available in early 1991.

Chapter 3

The Instruction Set

This chapter describes the instruction set of the transputer. As previously noted, the transputer cannot be seen to lie in either the classic RISC or CISC camps. For programmers used to other microprocessors it can seem strange. It has a small number of registers, organized as a stack, and all instructions are stack, rather than register, oriented. It has little concept of condition codes, there are limited instructions for accessing memory and no sophisticated memory-addressing modes.

On the other hand, the concepts of processes, inter-process communication and process scheduling are handled directly by the device. Fundamentally, multiple processor architectures and inter-processor communications are designed into the transputer and its instruction set.

This chapter provides an overview of the instruction set and does not intend to be exhaustive. A full list of the transputer instruction set is provided in Appendix A.

3.1 The evaluation stack

The instruction set of the transputer is designed around the idea of a stack as opposed to the idea of registers. Most other processors use these registers to manipulate values. Consider adding two numbers on a processor such as a 68000. The numbers would normally reside in memory and the result would eventually have to be stored back in memory. The 68000 is not capable of performing this directly even though it is a CISC processor. The sequence of operations would be as follows:

```
MOVE.L    memX,D0
ADD.L     memY,D0
MOVE.L    D0,memZ
```

which loads value `X` into the register `D0`, adds the value `Y` to it, then stores it back. In a RISC processor the `ADD` instruction would only work on registers, so the sequence could be described as follows (we continue to use 68000 style mnemonics for clarity):

```
MOVE.L      memX,D0
MOVE.L      memY,D1
ADD.L       D1,D0
MOVE.L      D0,memZ
```

Inside the transputer the approach is similar, except that a stack is used. The actual operations to add the two numbers would be as follows:

```
ldl         memX
ldl         memY
add
stl         memZ
```

Here the `ldl` instruction is used to transfer data from memory to a stack. The `add` instruction takes the top two values off the stack, adds them together and places the result on the top of the stack. The `stl` instruction is then used to store the top of the stack back into memory.

In a more conventional processor an assembly language programmer or compiler writer attempts to optimize performance by keeping suitable values in registers so that redundant access to memory is avoided. In the transputer the same approach is used to keep values within the evaluation stack as they are required. The stack is only three items deep and consists of the three registers A, B and C. The action of `ldl` can be described more formally as follows:

Copy the value of B into C
Copy the value of A into B
Load from memory into A

Thus two consecutive `ldl` instructions place two values from memory into A and B. The `add` instruction can again be described more formally as:

Add B to A
Copy the value of C into B
Leave C undefined

or informally as popping two values off the stack, adding them together and pushing the result onto the stack. In the same way the `stl` instruction pops a value off the stack and stores it into memory, or:

Store A in memory
Copy the value of B into A
Copy the value of C into B
Leave C undefined

In order to make effective use of this evaluation stack care must be taken in the order of evaluation of expressions, to ensure that wherever possible an intermediate expression is stored within the stack. The stack may seem limited in that it only holds three items but this is sufficient for most expressions.

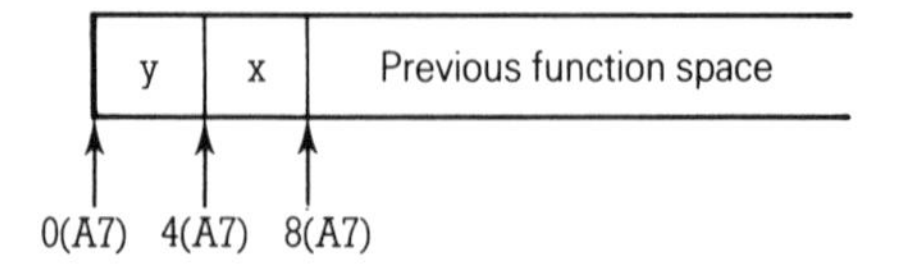

Figure 3.1 Local variable access in Motorola 68000 processor

3.2 Simple instructions

The previous chapter gave an introduction to the transputer registers. Explicit reference to a transputer register is never made, as access to registers is always implicit in each instruction. The instruction pointer *I* is manipulated by instructions which specify transfer of control while the workspace pointer *W* may be thought of as a stack pointer for a falling stack.

In a high-level language we need a way to store local variables and this is traditionally done with respect to a stack pointer. Consider the following fragment of C code:

```
f()
{
    int x, y;
    ...
    x += y;
    ...
}
```

Here two variables local to the function `f` are declared and used later in a calculation. Again it is worth considering how this is done in other processors. On a CISC machine such as a 68000 the address register `A7` is used as a falling stack. Space for local variables is allocated on this stack and can be addressed by positive offsets relative to `A7`. If we wish to store 32-bit values then these byte offsets must be multiples of 4, as shown in Figure 3.1.

Thus the code generated for the addition of `x` and `y` might look as follows:

```
MOVE.L      4(A7),D1
ADD.L       0(A7),D1
MOVE.L      D1,4(A7)
```

In the transputer, memory is allocated as offsets from the workspace pointer *W* (see Figure 3.2) in a very similar way, except that the offsets used are word offsets. The size of a word depends on the type of the transputer, being 16 bits for the T2 series and 32 bits for the others.

The instructions to load local data are always relative to *W* so it need not be specified. The only argument needs to be the offset from *W*, and so the same code fragment to add two local variables together would be generated as follows:

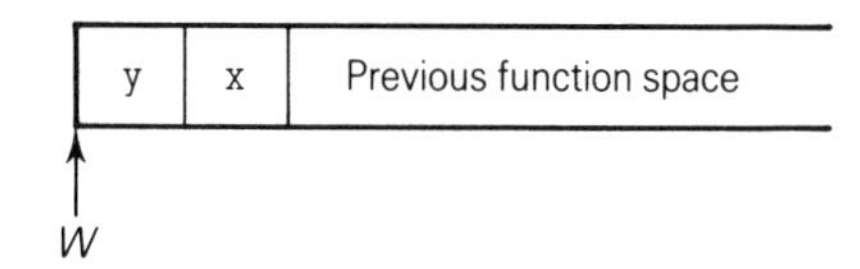

Figure 3.2 Local variable access in transputer

```
ldl        1
ldl        0
add
stl        1
```

This scheme has the advantage that the same code will work on a T212, a T414 or a T800 with 16-bit values used on the T212 and 32-bit values on the other processors. It also shows why 16-bit integers should not be used on 32-bit transputers as the only size of local variable supported is the word. There are ways of accessing a byte from memory, which are described below, but the only way to use 16-bit integers is to shift and mask 32-bit numbers. For this reason many compilers give a warning message if a high-level language program declares values as 16-bit objects.

Besides local variables in the workspace, access is also required to other areas in memory, commonly into structures in languages such as C. This is where indirection is required, using the value of one memory location to refer to another. Consider the following C code fragment:

```
main(argc, argv)
int argc;
char *argv[];
{
    char *arg;
    ...
    arg = argv[3];
    ...
}
```

Here the value of `argv` is a pointer to an array of argument slots, each of which contains a further pointer to a string. The assignment to `arg` causes an indirection via `argv` and the value of the index. The code to generate this on a 68000 might look as follows:

```
MOVEA.L    8(A7),A0
MOVE.L     12(A0),4(A7)
```

In the first line we access the array pointer stored on the stack and save it in `A0`. In the second line we indirect through an offset of 12 on `A0` to get the third array item, which is then stored as a local. Note that an offset of 8 is required because that is the byte offset of the third word in the array `argv`.

On the transputer we perform indirection using the `ldnl` instruction. The same code fragment would translate to transputer code as follows:

```
ldl     2
ldnl    3
stl     0
```

In the first line we access the pointer `argv` from the stack. The second line uses this plus the word given as argument as a pointer from which a memory location is loaded. The last line stores this back into memory again. Note that the transputer again uses a word rather than a byte offset for the argument to `ldnl`.

3.3 Encoding

The previous sections have given a very simple introduction to the transputer instruction set. We have seen how code fragments can be compiled on the transputer and have been able to compare this with a more familiar processor such as the 68000. This section describes the way in which the transputer encodes both instructions and data for greater code density.

Any computer architecture must choose a way in which numbers are stored. There are two differing requirements here. First, there is the maximum size of an integer which can be supported. On any 32-bit processor this is normally a 32-bit value. Secondly, there is the need to make the instruction set as compact as possible. For example, the 68000 series processors have an instruction to load an immediate value into a register. In the standard form this instruction is 6 bytes long: 2 bytes for the instruction code 'move immediate' plus 4 bytes for the value to be loaded. This is clearly wasteful when the number to be loaded is only small, and analysis of programs shows that loading small numbers such as 0, 1 or 4 is much more common than loading large numbers. The 68000 therefore has a special 'move quick' operation, which is only 2 bytes long and which can be used to load a number in the range −128 to +127.

There are several problems in having different instructions for the special case of small numbers. On the 68000 there are specific instructions for loading immediate values and adding or subtracting small numbers but the range is different for addition and subtraction, limiting these numbers to between 0 and 7. There are no special instructions for multiplication by small numbers or for any other immediate operations.

The transputer has a more elegant and flexible mechanism than this. A number within the transputer is represented by the number of bytes required and no more. The number is always represented in this way no matter where it is used within an instruction and no matter what the instruction. The approach is totally flexible giving no logical maximum integer size. A similar scheme is also used for encoding instructions, leading to a very dense and compact instruction set.

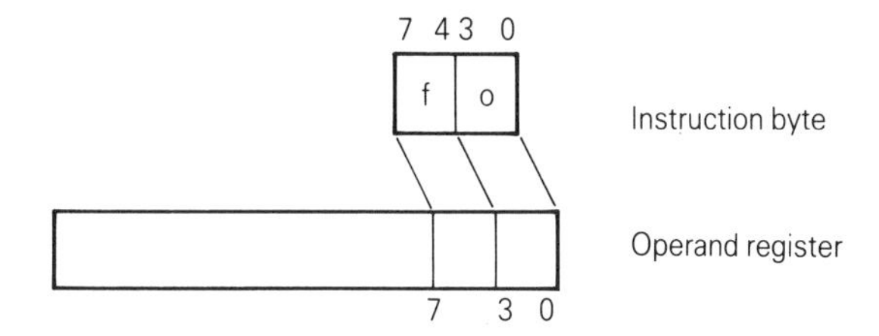

Figure 3.3 Instruction byte and operand register (f=instruction byte function (4 bits); o=instruction byte operand (4 bits))

Transputer instructions are encoded in a stream of bytes. Each byte consists of two nibbles, the upper four bits containing a function code, the lower four bits an argument (Figure 3.3). Thus there are sixteen different function codes, consisting of thirteen 'direct functions' and three special functions. For the thirteen direct functions the function code is the opcode of the instruction to be performed. Before executing the instruction the argument part of the byte is loaded into the bottom four bits of the operand register O, where it acts as the operand of the instruction.

Consider the encoding for loading a constant value onto the top of the stack. This instruction is called `ldc` for 'load constant'. This is a direct function and is represented by `0x4`. A number in the range 0–15 is represented by the argument nibble. Thus the complete instruction to load constant 8 onto the stack is represented by a single byte `0x48`. After the instruction has been executed, the operand register is cleared.

But often we will want to load a constant that is greater than 15. Larger operands are generated by using special functions `pfix (0x2)` and `nfix (0x6)`, which act as prefixes to the other functions. The `pfix` function loads its four argument bits into the operand register, and then shifts the register up four bits. The `nfix` instruction acts in exactly the same way, but complements the operand register before shifting it up. These functions are the only operations that do not clear the operand register after execution. For example, we can represent `ldc 0x15` by two bytes `0x21 0x45`, or `pfix 1; ldc 5`. The `pfix` instruction loads register O with the value `0x01` and then shifts it up four places, so that O contains `0x10`. The `ldc` instruction then inserts the argument from its own instruction word into the lowest four bits of O and uses the result `0x15` as the operand for the instruction. The register O is then zeroed ready for the next operation.

This mechanism is highly extensible and is limited only by the size of the internal registers. Table 3.1 shows the contents of the O register during the evaluation of the sequence of bytes required to represent the instruction `ldc 0x892`, which is `0x28 0x29 0x42`. In the same way `nfix (0x6)` may be used to construct a negative number. For example, the instruction `ldc -3` is represented by the byte sequence `nfix 0; ldc -3` or `0x60 0x43`.

Notice that it is also possible to encode numbers in a way that uses a fixed

Table 3.1 Execution of `ldc 0x892`

O before	Opcode	O during	O after
0	28	8	80
80	29	89	890
890	42	892	0

amount of memory simply by using `pfix` instructions with a zero argument. Hence `0x20` can sometimes be used as a no-op within the transputer, and it is useful to pad short numbers to a standard length with this code. This technique is sometimes employed when a linker is to be used to pad values after a compilation has taken place. A later section of this chapter discusses this problem in more detail.

The 'operate' function `opr` is used to extend the instruction set. The operand of `opr` is taken as the opcode to be performed. Thus:

```
opr    5
```

instructs the transputer to perform the opcode `0x5`, which is in fact `add`. The encoding for `opr` is `0xF` so the complete `add` instruction is encoded as `0xF5`. This scheme allows a further sixteen instructions to be encoded as a byte along with the thirteen encoded as a nibble. Again, these sixteen 1-byte instructions have been chosen to be the most commonly used after the direct functions. As the opcode of the instruction executed by the `opr` function is contained in the operand register, this register clearly cannot be used as the operand of the instruction. Thus instructions executed by `opr` take their operands from the evaluation stack or from memory.

Further instructions are available using larger operands to the `opr` instruction. For example:

```
opr    0x41
```

executes opcode `0x41`, which is a shift left instruction. This is encoded as `pfix 4; opr 1` or `0x24 0xF1`, using the `pfix` instruction to extend the number range of the opcode.

This may sound rather complicated but the entire mechanism can be encapsulated in a few lines of code. The following C routine is used to encode a number in the most efficient manner:

```
#define f_pfix 0x20
#define f_nfix 0x60
#define TRUE  1
#define FALSE 0

void encodestep();

void encode( op, arg)
int op, arg;
```

```
{
   if (arg < 0)
      encodestep((~arg) >> 4, TRUE);
   else
      if (arg > 15) encodestep(arg>>4, FALSE);
   printf("%2x\n", op | (arg & 0xf));
}

void encodestep(arg, negative)
int arg, negative;
{
   if (arg > 15)
   {
      encodestep((arg>>4), negative);
      printf("%2x", f_pfix |
            ((negative ? ~arg : arg) & 0xf));
   }
   else
      printf("%2x", (negative ? f_nfix : f_pfix)
                    | (arg & 0xf));
}
```

Here are some encodings of various numbers:

```
ldc    12           0x4c
ldc    40           0x2248
ldc   257           0x212041
ldc   -12           0x6044
ldc   -40           0x6248
ldc  -257           0x612F4F
```

Clearly the encoding scheme can result in a very compact instruction stream. Reverting back to the sample segment of C code to add together two local variables, we can compare the size of the instruction set on the CISC processor and the transputer. The 68000 version is as follows:

```
MOVE.L     4(A7),D0       0x202F0004
ADD.L      0(A7),D1       0xD2AF0000
MOVE.L     D1,4(A7)       0x2F410004
```

Compare the size of the code with the transputer version, which is as follows:

```
ldl        1             0x71
ldl        0             0x70
add                      0xF5
stl        1             0xD1
```

The thirteen instructions allocated to single nibbles have been carefully chosen to be those used most commonly and are shown in Table 3.2. These include loading a constant, a local variable, a non-local variable, a local pointer and a non-local pointer. It is also possible to store local and non-local variables, adjust the W pointer register, add and compare constants, make conditional and non-conditional jumps and call procedures using a single nibble. Sixteen more of the commonly used instructions are encoded as single bytes (Table 3.3). The 'Cycles' column shows the number of processor clock cycles taken to execute each instruction when its non-register operands are in internal RAM. Many of these simple instructions execute in a single clock cycle.

Table 3.2 Direct functions

Function code	Mnemonic	Cycles	Instruction
0x0	j	3	jump
0x1	ldlp	1	load local pointer
0x2	pfix	1	prefix
0x3	ldnl	2	load non-local
0x4	ldc	1	load constant
0x5	ldnlp	1	load non-local pointer
0x6	nfix	1	negative prefix
0x7	ldl	2	load local
0x8	adc	1	add constant
0x9	call	7	call
0xA	cj	2/4	conditional jump (taken/not taken)
0xB	ajw	1	adjust workspace
0xC	eqc	2	equals constant
0xD	stl	1	store local
0xE	stnl	2	store non-local
0xF	opr	–	operate

Table 3.3 Single-byte operations

Byte	Mnemonic	Cycles	Instruction
0xF0	rev	1	reverse
0xF1	lb	5	load byte
0xF2	bsub	1	byte subscript
0xF3	endp	13	end process
0xF4	diff	1	difference
0xF5	add	1	add
0xF6	gcall	4	general call
0xF7	in	2w+19	input message
0xF8	prod	x	product
0xF9	gt	2	greater than
0xFA	wsub	2	word subscript
0xFB	out	2w+19	output message
0xFC	sub	1	subtract
0xFD	startp	12	start process
0xFE	outbyte	23	output byte
0xFF	outword	23	output word

w= number of words transferred.

3.4 Further instructions

Many of the instructions in the transputer are easily recognizable in other processors and are not worth an explicit mention here. A full list of the transputer instruction set is given in Appendix A.

These instructions include standard operations, such as signed and unsigned arithmetic operations; logic operations such as `AND`, `OR`, `EXCLUSIVE-OR`; and shifts. Unlike many other microprocessors, the transputer does not contain a special piece of hardware called a barrel shifter which is used to implement shifting of data. This means that shifting data is slow compared to some other processors; in particular, the unchecked multiply instruction called `prod` should be used in preference to shifts when multiplying by a constant power of two.

The transputer also contains explicit support for operations on values longer than a word; these are especially useful on the 16-bit series. Versions of the signed and unsigned arithmetic operations are provided, as well as extended shift instructions to shift double-sized objects. The normalize instruction `norm` may be used to shift left a double-sized object until its most significant bit is one, keeping track of the number of shifts needed.

As mentioned earlier, the transputer normally works entirely in words, which are either 16 or 32 bits depending on the type of transputer. It is possible to access individual bytes through instructions that allow loading and storing bytes. There is also provision for handling byte-sized structures, so that the same code could work on transputers with different word sizes. The `bsub` instruction is used to return the address of an offset within a byte structure. In the same way the `wsub` instruction returns the address of an offset within a word-sized structure, avoiding the need for explicit multiplication by two or four. The `bcnt` instruction can be used to multiply the A register by the number of bytes in a word, whatever that is. A complementary instruction `wcnt` decomposes an address into a word offset and a byte selector.

Structures can be copied efficiently using the `move` instruction, which takes its arguments from the evaluation stack. This instruction will copy A bytes of data from the address in C to the destination address in B. This is an efficient one-dimensional copy and the more recent transputer variants, such as the T800 and T425, also have two-dimensional copy instructions. The `move` instruction does not allow overlapping source and destinations.

3.5 Position-independent code

An important feature of many modern microprocessors is that code produced for them can be 'position independent' in that the code contains no absolute addresses. The instructions may be loaded anywhere in memory without the need for any changes to the code. This is normally good programming practice as it is also a prerequisite for producing 'pure code', which can be shared by many different processes. It also allows for code to be moved once it is loaded into memory.

On the transputer the use of pure code is highly important. As the transputer supports multiple processes in the instruction set it is important that standard libraries be shared by all the processes. Position-independent code may also be passed between interconnected processors, being loaded into different memory locations in different processors. The production of position-independent code is made possible by the `ldpi` instruction, which loads a pointer to an instruction. Consider attempting to load the address of a string, ready to print it out with `printf` or some such similar routine. This string will normally be located somewhere within the program code. The following code would be used to perform the required action:

```
        ldc (L1-L2)
        ldpi
L1:      ...
        ...
L2:      byte "Hello World"
```

Here the A register is loaded with the offset of the string from the instruction following the `ldpi`, which is used to add in the current value of the instruction pointer or I register. The sequence `ldc 0; ldpi` may be used to access the address of the next instruction.

3.6 Flow of control

The transputer supports the normal range of instructions to alter the flow of control and these are listed in Table 3.4. The `call` instruction is used to call a procedure; the instruction pointer I is pushed onto the stack, followed by the three registers A, B and C. The argument to `call` is then added to I to transfer control. The new

Table 3.4 Flow of control instructions

Code	Mnemonic	Cycles	Description
Jumps			
0x0	j	3	jump – direction function
0xA	cj	4	conditional jump (taken)
		2	(not taken) – direct function
0x21	lend	10/5	loop end (loop back/exit)
Call and return			
0x9	call	7	call – direct function
0x06	gcall	4	general call
0x20	ret	5	return

function will then commonly allocate some workspace by using the `ajw` instruction to decrement W. At the end of the routine the workspace used is deallocated with another `ajw`; the `ret` instruction is then used to deallocate the four slots of workspace used by `call` and to jump back to the place indicated by the saved I. This mechanism allows up to three arguments to be passed in the registers; extra arguments would normally be pushed onto the workspace stack before the call.

The `call` instruction is normally useful for calling routines where the offset is known in advance. This is not the case when calling a routine where the address is to be computed at run time. The instruction `gcall` is provided which simply swaps I and A, thus enabling the address to be computed into A and then jumped to. Another `gcall` could be used to return.

It is also often useful to be able to write a routine in such a way that it can be called from either `call` or `gcall`. The easiest way to do this is to assume that all routines are going to be accessed via `call` instructions. If a `gcall` must be used, a 'stub' procedure is generated and a `call` is made to this instead, passing the address of the procedure as an extra argument. The call to the stub will set up the workspace correctly and the stub needs only to pick up the extra argument at the end and jump to this using `gcall`.

Transfer of control can also be made using unconditional and conditional jumps `j` and `cj` while for-loop control is handled with `lend`. The transputer programmer should note that the instructions `j` and `lend` both have a curious side-effect – they may destroy the contents of all the evaluation stack registers A, B and C. This can be very confusing because at first it seems slightly odd that the registers are trashed at these points, and secondly because they are not always altered. So an assembler routine might work at some times and not at others.

3.7 Process scheduling

The curious side-effect of registers *sometimes* being altered, mentioned above, is easy to explain. The transputer instruction set supports multiple processes, and the low-priority processes are time sliced by the chip. This means that once one low-priority process has used a certain amount of CPU time, the processor switches to another low-priority process waiting to run. This switch is a hidden transfer of control that in theory can take place at any time. However, the designers of the transputer wanted to make the process switch highly efficient. In order to make a process switch the chip has to save the current I pointer in the workspace, save the workspace pointer W somewhere in memory, retrieve another workspace pointer W' for a new process, retrieve the saved I' from W' and start execution again. If this was to take place at any time the microcode would need to also save A, B and C from the original process, and reload A', B' and C' for the new one. This is in fact what is done when a high-priority process interrupts a low-priority one; but because there is only one high-priority process queue, and high-priority processors never interrupt each other, the registers may be saved in fixed locations in fast on-chip RAM. For low-priority processes it would mean six extra external memory accesses on each time slice if the registers had to be saved in this way.

The solution instead is only to attempt to switch between low-priority processes at certain points, and not to bother to save *A*, *B* and *C* when making the switch. The only question is: at what point in a program can switching be allowed? Clearly, trashing *A* between a `ldl` and the next `adc` is no good at all, so the chip must attempt to switch between processes only on certain instructions. The instructions chosen must appear regularly in all programs so that time slicing always takes place. The obvious choices are the instructions that transfer control, because they must appear in a program that is looping and therefore taking CPU time. Thus at each `j` and `lend` instruction the processor performs its time slice test and might switch to another process. Process switching can also take place when processes are waiting for communication; this is described in more detail later.

It is important to be aware of the underlying time slicing when programming the transputer in its assembly language. First, compiler writers should ensure that their compiler code includes the loop and jump instructions, which is of course normally the case anyway. Secondly, any hand-coded assembler that is attempting to perform something time-critical should not include `j` or `lend` if running at low priority because both instructions have the possibility to take a large amount of real time if other processes are scheduled at that point. Finally, two processes might wish to communicate using shared data and it is important that the data remain in a consistent state. The programmer might wish to ensure that no time slicing takes place between the start and end of a certain operation. In this case `lend` should be avoided and conditional jumps on false (`ldc 0`); `cj`) used instead of `j`. This puts an extra zero into the top of the stack which can be removed by an add instruction at the destination of the jump.

The processes that are ready to run, and which are time sliced if they are at low priority, are held internally in two lists, one for each priority. The next process to run is held at the head of the list. The `startp` instruction can be used to add a new process to the correct list and `endp` used to terminate the current process. The `endp` instruction is slightly more complicated as it is passed the identity of the process to run next, and this successor process is not actually started until a specified number of processes have all executed `endp`. The effect is that one process can create a number of child parallel processes, perform some work and then execute `endp` itself. When all of the child processes have completed, the parent process is reactivated.

A process can ask to suspend itself using the `stopp` instruction. It remains in a state of suspended animation until another process executes a `runp` instruction on it. The value passed to `runp` specifies the process identity and also whether it is to be restarted at high or low priority. A process can determine its own priority using the `lpri` instruction.

3.8 Channels

Communication between transputer processes is handled by what are called 'channels'. Data may be sent down a channel by one process and received by another process waiting on that channel. If there is no process waiting to receive the data from the channel, then the sender waits. The concept of the channel is central to the philosophy of the transputer, because a channel is used to transmit information not only between processes in the same processor but also between processes on different processors. In each case the instructions used are the same, although in the internal version the data are copied in memory, while in the external version the data go out on one of the serial links. This means that a piece of compiled code can be passed channels as arguments and the code need not distinguish between internal and external cases.

A channel is controlled by a word in memory. This holds either a process identifier which represents a process and its priority, or the most negative number `MinInt`, which cannot represent a process. Eight memory locations at the low end of the address space are reserved as channel control words for the external channels connected to the duplex serial links. Any other memory location may be used as an internal channel control word, but it must first be initialized to the most negative integer using the `mint` instruction.

Communication is achieved using the instructions `in`, `out`, `outbyte` and `outword` (see Table 3.5). The `out` instruction takes a message pointer, a length and a channel control word address. It transmits the information via the channel to a process executing the `in` instruction. This states how many bytes are expected, from which channel, and where the incoming buffer is located. The `outbyte` and `outword` instructions can be used to transfer a single word or byte to a specified channel; there are no similar input instructions.

Channel control words are initialized with `MinInt`. The first process to be ready to use the channel executes, say, an `in` instruction. This looks at the channel control word, notices that it is initialized and hence detects that this is the start of the transaction. The location of the message buffer is saved at a negative offset on the workspace stack. The value of the channel control word is then altered to the

Table 3.5 Channel communication and associated instructions

Code	Mnemonic	No. of cycles	Description
0x0E	outbyte	23	output byte
0x0F	outword	23	output word
0x07	in	2w+18	input message – proceeds
		20	– communication waits
0x0B	out	2w+20	output message – proceeds
		20	– communication waits
0x42	mint	1	load *A* with MinInt
0x12	resetch	3	reset channel

The code shown is the operand required by the `opr` function which executes the instruction.

current process identifier and the current process is descheduled.

When the second process executes the other half of the transaction the `out` instruction will detect that the channel has another process waiting on it. It retrieves the receive message buffer saved in the workspace of the waiting process and copies the number of bytes specified in the `out` to perform the data transfer. It then resets the channel control word back to `MinInt` to mark it empty and reschedules the waiting process.

If the processes happened to execute in the other order then the procedure would be similar. The `out` instruction would save the output buffer, set the channel control word to the current process and deschedule the process. The `in` instruction would then actually perform the copy, using the argument specified to `in` to determine the length of the transfer.

From this description it will be clear that messages sent down channels must be received using the same length `in` as was used with the `out`. Similarly, an `in` specifying a message buffer of 1-byte or 1 word must be used to receive data sent with `outbyte` and `outword`. It is not possible to receive, say, 5 bytes sent as a 4-byte `outword` and a 1-byte `outbyte`. This is because the number of bytes actually transferred would depend on which process executed its input or output instruction last, and hence whether the number of bytes to be transferred was taken from the `in` or the `out` instruction.

External channels are implemented using a DMA transfer down the link. The process starting the transfer is descheduled until the transfer has completed. When the receiving process in the other transputer executes the other half of the transaction, that process too is descheduled while the data is handshaked over the links. When both processors have written and read the number of bytes specified to the `in` and `out` instructions, each process involved is rescheduled. This means that it is in fact possible to send a message down a link using several `out` instructions and receive it with a single `in` instruction. Using this aspect of the transfer would mean that a particular program would need to be aware of the type of channel in use. The instruction set has been carefully designed so that this is not normally necessary.

Some care needs to be taken when exchanging messages down links. If the message is a single transfer then there is no problem, but the message must be of fixed length so that both processes know how many bytes are to be written and read. More commonly, variable length messages would be required, and in this case the message could have a fixed length header containing the length of the following variable length data section. Again, there is no problem if there is only one process performing output on a particular channel, as the header will always be sent, followed by the data section.

Problems can start to occur if more than one process is allowed to output a message, as can happen when different processes are attempting to print debugging information, for example. In this case the first process can output a header, and then become descheduled. A second process might then run which also starts to output a header, but sends this header when the receiving process is expecting the variable length data section of the first message. The result is a pair of garbled messages.

There are two solutions to this problem. The first is always to have a sender process to which messages are sent; only this process is allowed to perform output down the link. This mechanism works reasonably well but the message may have to be copied twice unless the sender process is allowed read access to any data held in memory. This is of course always physically the case, but in many software systems internal message passing by copying rather than by reference is preferable because the reference mechanism does not work between different processors.

The second solution is to make use of a semaphore flag, which is used to guard the use of the link. If a process wishes to send a message it sets the flag, which is inspected by other processes wishing to use the link. The second process can either go to sleep for a while, or it can add its message to a queue that is handled by the first process just before the semaphore is unset.

Channels can be reset using the `resetch` instruction. This is useful when communication between two processors has failed for some reason and the channel is to be brought back into use again. The result of `resetch` is the value stored in the channel control word. If this is a process identifier then it is possible to restart the process waiting on the channel using the `runp` instruction. It would also be necessary in this case to ensure that the process was informed that it had been rescheduled in this way and that the data transfer had not taken place as expected.

3.9 Timer

Like many microprocessors intended for embedded applications the transputer contains two internal timers. One is associated with high-priority processes and has a 'tick' every microsecond. The other is associated with low-priority processes and has a 'tick' every 64 μsec. The high-priority clock is used for, among other things, the time slicing between low-priority processes. Every 1024 ticks the current process is examined. If it is the same process as was running at the last inspection 1024 μsec ago then it is descheduled at the next `j` or `lend` instruction as described above. This means that low-priority processes are time sliced at irregular intervals greater than 1024 μsec.

The current value of the clock can be read using the `ldtimer` instruction. The actual clock read depends on the priority of the process doing the reading. The value is returned in ticks, so the process priority needs to be taken into account when comparing the number of ticks with real time. The clock ticks are also only held as a single 32-bit register (or 16-bit) and so wrap around after a certain amount of time.

A process may be put to sleep for a specific amount of time by specifying a timer input using the instruction `tin`. This instruction causes the process to become descheduled until the specified time has passed. The process will become ready to run again after the time has elapsed; it may not actually run until a little after as it may have to wait for the current process to be time sliced. The following code fragment will cause a low-priority process to be delayed for a second or a little more. Note that 15,625 is the number of 64 μsec ticks in one second.

```
ldtimer
ldc 15625
sum
tin
```

The value returned from `ldtimer` is unsigned, so it is important to use the unsigned operators `sum` and `diff` rather than `add` and `sub`. Such a process could be used, for example, to maintain a real time clock by adding to a seconds, minutes and hours field when it was reawakened.

3.10 Alternation

The section above described the simple use of the clock. A more common use of a timer is to time out events that might fail, especially timing out messages sent via external channels to other transputers and which may fail to get there. In this case we want to be able to say that a process is to wait until either the message completes or a certain timeout period has expired. Thus the transputer instruction set supports alternation, the ability to choose which of a number of processes to schedule, according to some condition occurring outside the processes: a channel input becoming ready, a timer input or an `Event` input.

The basic design of the alternate implementation is as follows (see Table 3.6). The `alt` instruction is executed first. This is then followed by a number of enable instructions that identify the different alternative events to be waited for. The `altwt` instruction is then executed and this causes the process to be descheduled until one or more of the alternative events occurs. A matching set of disable instructions are then executed, one for each event that was enabled. The order in

Table 3.6 Alternation instructions

Code	Mnemonic	No. of cycles	Description
0x43	`alt`	2	alt start
0x44	`altwt`	5	alt wait – channel ready
		17	alt wait – channel not ready
0x45	`altend`	4	alt end
0x49	`enbs`	3	enable skip
0x30	`diss`	4	disable skip
0x48	`enbc`	7	enable channel – ready
		5	enable channel – not ready
0x2F	`disc`	8	disable channel

which the disable instructions are executed is used to specify a priority of items where more than one event has occurred. The result from the disable indicates whether that event actually happened. Finally an `altend` instruction is executed, which causes control to be transferred to the correct code to handle the event that happened.

The enable instructions specifying alternatives are used to construct language elements called 'guards.' A guard has a boolean part, which is constructed as normal, and then either a check for a channel, a timer or nothing at all. Each of these three possibilities has its own instruction – `enbc`, `enbt` and `enbs` respectively. They all take the boolean part of the guard in A and the channel or timeout in B. Unless the boolean is true the enable instruction has no effect.

After the `altwt` instruction has been executed a matching set of disable instructions from the set `disc`, `dist` and `diss` must be executed. These must be passed the same booleans and values as their enable counterparts; they must also be passed the offset of an address which contains the code used to handle that alternative. Once the `altend` instruction is executed control passes to this address. If the alternative involved waiting for a channel then the first action should be to read that channel. The disable instructions themselves also return `TRUE`; this can be used to distinguish different alternatives where the same code address is used for each.

The following example would wait for one of two channels, and read the data from the first:

```
        alt              - start alt
        ldl   1          - load channel address
        ldc   1          - load boolean true
        enbc             - enable channel 0
        ldl   2          - load channel address
        ldc   1          - load boolean true
        enbc             - enable channel 1
        altwt            - wait for one event
        ldl   1          - load channel address
        ldc   1          - load boolean true
        ldc   L0-LA      - offset of L0 from altend
        disc             - disable channel 0
        ldl   2          - load channel address
        ldc   1          - load boolean true
        ldc   L1-LA      - offset of L1 from altend
        disc             - disable channel 1
LA:  altend
- Case channel 0 ready
L0:  ...
- Case channel 1 ready
L1:  ...
```

If any of the alternatives include waiting for the timer via an `enbt` or `dist`, then the `alt` and `altwt` instructions must be replaced by `talt` and `taltwt`.

3.11 Error handling

Unlike many other microprocessors the transputer supports only one type of error in its integer processor. The value of an internal flag called *ErrorFlag* is normally set to zero; under certain conditions this may be changed to one. Unfortunately these conditions do not include attempting to access a non-existent memory location, or accessing a word from an address that is not a multiple of the number of bytes per word. Both of these common errors are not detected at all.

The *ErrorFlag* can be set explicitly by using the `seterr` instruction (see Table 3.7). It can be cleared by the `testerr` instruction, which loads A with true if the *ErrorFlag* was set. The state of the *ErrorFlag* is reflected by an output pin `Error`. The pin can be connected to some external hardware, which can cause this state to be detected by a host processor, or which can turn on an error lamp.

The status of the *ErrorFlag* can be checked with the `testerr` instruction; normally a sequence of operations to be checked would be enclosed by `testerr` instructions – the first to clear the flag and the second to read it. A suitable error message could then be given if the *ErrorFlag* was set. The `stoperr` instruction can be used instead of the final `testerr`; this stops the current process (as in `stopp`) if the *ErrorFlag* is set.

Alternatively, the transputer can be placed in a mode where the processor is brought to an immediate halt if *ErrorFlag* is set. This action is controlled by the *HaltOnErrorFlag*, and instructions are provided to clear, set and test this.

The *ErrorFlag* is always set when arithmetic overflow occurs, or when the transputer executes a specific error check instruction that fails. Arithmetic overflow can occur on any of the following instructions:

`adc` add constant
`add` add
`sub` subtract

Table 3.7 Error-handling instructions

Code	Mnemonic	Instruction
0x29	`testerr`	test *ErrorFlag*false and clear
0x10	`seterr`	set *ErrorFlag*
0x55	`stoperr`	stop on error
0x57	`clrhalterr`	clear *HaltOnError*
0x58	`sethalterr`	set *HaltOnError*
0x59	`testhalterr`	test *HaltOnError*
0x13	`csub0`	check subscript from 0
0x4D	`ccnt1`	check count from 1

`mul`	multiply
`div`	divide
`rem`	remainder
`ladd`	long add
`lsub`	long subtraction
`ldiv`	long divide
`fmul`	fractional multiply

The error check instructions may be used to check that parameters passed to various operations are legal. The `csub0` instruction sets *ErrorFlag* if B is greater than A and can thus be used to check subscripts. The `ccnt1` instruction is similar, except that it checks that the value in B is greater than zero and less than or equal to A.

Although the transputer normally deals with words, either 16 or 32 bits long, support is provided to extend part words such as bytes to words and to extend words to double words. These conversions are achieved by using the `xword` and `xdble` instructions. Two matching instructions `cword` and `cdble` are used to detect whether a word can be represented in a part word, or whether a double can be represented by a word. If the value cannot be represented in this way the *ErrorFlag* is set.

3.12 The extended instruction set

The instructions detailed above have applied to all members of the transputer family, be they 16- or 32-bit versions. However, the introduction of the T800 saw some extra functionality included. Although the majority of the extra T800 instructions refer to on-chip floating-point support described in a later section, the T800 also had a number of extra general-purpose instructions added (see Table 3.8). This extended instruction set has been recently included in the T222, T225, T400, T425, T801 and T805 variants.

The most elementary extra instruction is `dup`, which simply duplicates the top of the stack, moving A to B and B to C. This useful instruction was, one assumes, simply forgotten in the original design of the T414.

In addition to `dup`, another extra instruction `wsubdb` is provided to aid the manipulation of double-word quantities. This behaves exactly like `wsub` except that each item is assumed to be two words rather than one word long.

The next section of support provided in the later chips includes more instructions to handle bit manipulation. These are useful in various graphics routines and also in the computation of cyclic redundancy checks or CRCs. The instruction `bitcnt` may be used to count the number of bits set to one in a word. It adds the total number of bits set in A to the count in B, with the final result appearing in A. This is useful in a loop counting the number of bits set in an array. The instructions

Table 3.8 Extended instruction set

Code	Mnemonic	Instruction
0x5A	dup	duplicate top of stack
0x5B	move2dinit	initialize data for two-dimensional block move
0x5C	move2dall	two-dimensional block copy
0x5D	move2dnonzero	two-dimensional block copy non-zero bytes
0x5E	move2dzero	two-dimensional block copy zero bytes
0x74	crcword	calculate CRC on word
0x75	crcbyte	calculate CRC on byte
0x76	bitcnt	count bits set in word
0x77	bitrevword	reverse bits in word
0x78	bitrevnbits	reverse bottom n bits of word
0x81	wsubdb	form double-word subscript

`bitrevword` and `bitrevnbits` are used to reverse either all or part of the bits in a word.

Support for CRC calculation is provided via the `crcword` and `crcbyte` instructions, which are designed to be used in an iterative loop, once per word or byte in the array whose CRC is to be calculated. The A register holds the word or byte that is to be combined with the accumulated CRC held in B, using the generator in C.

3.13 Graphics support

Extra instructions are also provided in later variants of the transputer for graphics applications. These center on the concept of two-dimensional block moves. The standard instruction set includes the `move` instruction, which copies a certain number of bytes in memory from one point to another. This can be thought of as copying a single row of memory, while the two-dimensional copy can be viewed as copying a rectangular area.

Consider the diagram in Figure 3.4. The set of bytes representing the rectangle is actually stored in memory as three rows of ten consecutive bytes. In order to describe this area, we must know the start address of the first row, the width of each row and the number of rows to be copied. We can also refer to the number of rows as the length of the rectangle. In addition we need to know the stride being used, which is the number of bytes to be added to the address of the start of the first row to get to the start of the second row. If the stride is equal to the width then all the bytes are consecutive in memory.

It is clearly useful to be able to specify a stride greater than the width. Consider a video screen in memory providing a display of 512 rows by 512 columns, with

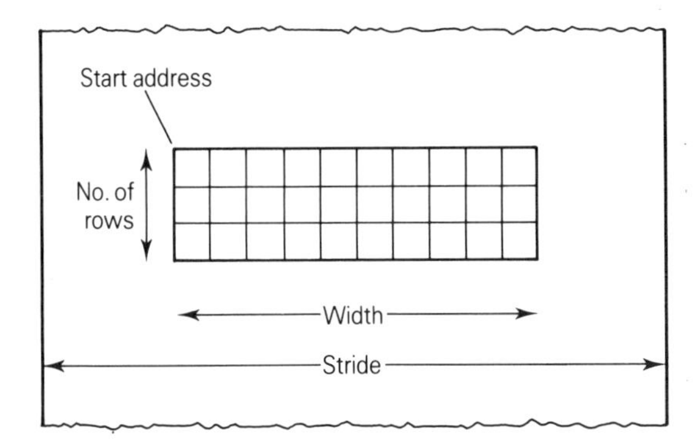

Figure 3.4 A rectangle in memory

each pixel being a single byte of video memory. We shall assume that the screen memory is contiguous through the address space. The screen can be viewed as a two-dimensional array, so that in C the top left pixel would be `screen[0,0]`, the top right pixel would be `screen[0,511]` and so on. Alternatively we could regard the screen as a one-dimensional array; in this case `array[511]` would refer to the top row, rightmost pixel while `array[512]` would refer to row 1, leftmost pixel.

Within this screen we wish to be able to copy a rectangle representing a window from one place on the screen to another. In the following diagram we assume that the window has its top left corner at (2,10) and is of width 100 pixels and height 50 pixels (Figure 3.5).

The parameters needed in this case to describe the window are as follows:

- The base address. This is the address of the first pixel in the window which is at offset `(row*stride)+column` from the start of the screen or `&array+1034`.
- The width of the rectangle, which in this case is 100.
- The length of the rectangle, which is 50 in our example.
- The stride, which is the distance from the start of one row to the start of the next, or 512.

With this information we are ready to use the two-dimensional copy in its simplest form. We provide the description of the source rectangle, and need only to supply the address of the destination. The instruction to perform the move is called `move2dall`, but it requires more than three arguments. This problem is handled by performing the instruction `move2dinit` before the actual `move2dall`, to preload the first set of parameters. The instruction sequence would be of the following form:

```
ldc    512          - load source stride
dup                 - load destination stride
ldc    50           - load length
move2dinit          - set up first parameters
```

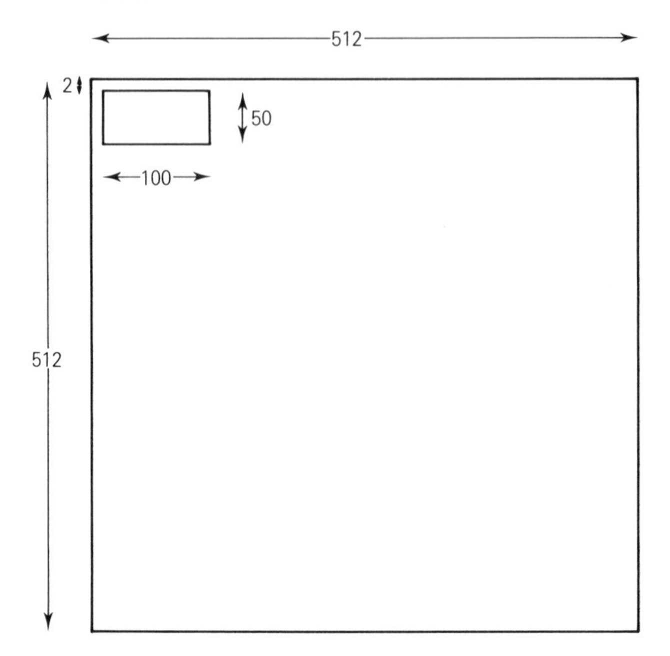

Figure 3.5 A screen with a rectangular window

```
ldc     1034        - load source offset
ldl     arrayptr
bsub                - Ptr to source
ldc     411         - load destination offset
ldl     arrayptr
bsub                - Ptr to destination
ldc     100         - load width
move2dall           - perform the copy
```

The final `move2dall` will copy all the bytes in the rectangle to their new position, in one instruction. This is extremely fast, and can be used to move rectangular areas around within the video flyback time.

The two-dimensional copy can be used in many more ways than it might at first appear. Note that in the code example above a `dup` instruction was used to set the destination stride to the same as the source stride. In general this need not be the case, and it is very useful to be able to make these values different. For example, we may have wanted to copy the window off the screen into some area which was not part of the video RAM, and hence not visible. In this case we would want to represent the window in the most compact way possible, with rows of the window laid end to end. We would perform the same instruction sequence but with the destination stride set to equal the width of the rectangle. We could then replace the window on the screen with the source and destination strides reversed.

The two-dimensional copy can also be used to clear a screen, by setting all the bytes to a particular color. First we take a single byte of memory and set this to the required value. Next we set up a two-dimensional copy, providing the address of this byte as the starting point. We specify the width as one, as we have only a single byte in our source.

We then specify the length of the source area as the number of pixels in our destination; for example, the entire screen would require a value of 512 × 512. The trick is to set the stride of the source area to zero, while setting the stride of the destination area to one. The `move2dall` instruction will perform a row copy from the source to the destination while counting down the value given as length. At the end of each row, specified by the value for width, the source stride is added to the source row start address and the destination stride added to the destination row start address.

In this case the source stride is zero and the destination stride is one; thus the single byte is copied into a large contiguous area of memory in one instruction.

The technique can be modified to clear a window on the screen. First, a single byte is copied into the first row of the window using a single `moved2all` as described above. A further two-dimensional copy is then set up which specifies this first row as the source, and the start of the second row of the window as the destination. The width is given as the width of the window, while the length is set to the number of rows in the window less the one already done. The source stride is again set to zero, while the destination stride is set to the stride for the screen, 512 in our example.

The `move2dall` instruction actually copies data from the source to the destination area. There are two further instructions, which behave in a slightly different fashion. The `move2dnonzero` instruction will only copy bytes that are non-zero. The bytes in the destination corresponding to zero bytes in the source are left unchanged. This may be used to copy an outline onto a picture already present in the destination.

The `move2dzero` copies only zero bytes to the destination. This can be used to zero selectively certain parts of the destination according to a template provided as source.

3.14 Floating-point support

The original T414 transputer had little support for floating point. A number of instructions were provided in order to aid the implementation of floating-point arithmetic in software, but there were no floating-point instructions implemented in hardware. This is in common with most microprocessors, where floating-point support is often added by means of an external coprocessor.

The T4 series processors provide the instruction `unpacksn` which turns an IEEE format floating-point number into the constituent exponent and mantissa parts. This operation is common to all the code to emulate floating point in software.

Similarly the instructions `roundsn`, `postnormsn` and `norm` are used to normalize, round and then pack the expanded exponent and mantissa back into IEEE format.

The final part of the T4 support package includes `ldinf` to load the IEEE representation of infinity, and `cflerr` which sets the error flag if the *A* register contains infinity or the IEEE representation of 'not a number'. Table 3.9 shows the T4 floating-point support instructions.

The T8 series provides a full implementation of IEEE-754 32-bit and 64-bit floating-point arithmetic in hardware, and as a result it does not include the instructions described above. This is a nuisance as it makes it difficult to provide floating-point code which works on either a T4 or a T8. Clearly, any floating-point intensive code should be targeted for the T8, as the instructions for floating point are extremely fast. This is because the floating-point unit is internal on the T8 series devices, and is not an external coprocessor, unlike many other microprocessor systems.

The speed of a T8 is also enhanced by the asynchronous nature of the floating-point unit, allowing integer operations to continue while the floating-point unit is busy. This is of particular importance where an address must be calculated, such as the address of an element in a multi-dimensional FORTRAN array. In a code fragment such as `A(I+J) = X *Y` the floating-point multiplication should be set running as soon as possible. The floating-point operations use an extra stack of registers called *FA, FB* and *FC*. Where possible, the values of `X` and `Y` should be loaded in registers *FA* and *FB*, and then the floating-point multiply instruction `fpmul` executed. This will return a result into *FA*, but while the floating-point processor is running and not interacting with the integer registers, the integer engine is available for work. The instructions to compute the address of `A(I+J)` will execute in parallel with the floating point multiplication. The integer processor and the floating-point processor resynchronize when the value of *FA* is stored to memory.

Many of the extra T8 instructions use a standard instruction called `fpentry`, which causes a specific floating-point instruction to be executed depending on the value of *A*. These routines have mnemonics which start `fpu`, so that, for example, `fpuabs` replaces register *FA* with its absolute value. This is actually implemented by performing the following two instructions:

Table 3.9 T4 floating-point support instructions

Code	Mnemonic	Description
0x63	`unpacksn`	unpack single-length number
0x6D	`roundsn`	round single-length floating-point number
0x6C	`postnormsn`	post-normalize correction of single-length floating-point number
0x71	`ldinf`	load single-length infinity
0x73	`cflerr`	set error flag if A value is either floating-point infinity or Not-a-Number in IEEE representation

```
ldc fpuabs_value
fpentry
```

although some assemblers do this work automatically.

The instruction set additions for floating point are not described in any more detail here, except to say that besides supporting floating-point arithmetic for single and double precision numbers, complete with error checking where required, they also provide hardware support for floating-point remainder and square root. These last two operations are particularly unusual as they are implemented as a sequence of instructions rather than a single opcode. The reason for this is that they are likely to take a long time, and the designers of the transputer did not wish to extend the potential interrupt latency period. Instead they provide a set of two or three instructions to perform the single job, allowing interrupts to occur between each instruction. Thus the code sequence to produce a single precision square root, rounded to the nearest representable value is as follows:

```
fpusqrtfirst\index{fpusqrtfirst}
fpusqrtstep\index{fpusqrtstep}
fpusqrtstep
fpusqrtlast\index{fpusqrtlast}
```

The instructions must be performed in this order, the `fpusqrtstep` instruction being repeated, in order to compute the square root. Any other combination is likely to crash the transputer.

The remainder calculation uses two instructions: `fpremfirst` which is executed once, and `fpremstep` which is executed repeatedly until it returns the value `0` to the *A* register. The code to compute a remainder is as follows:

```
        fpremfirst
        eqc 0
        cj next
loop:   fpremstep
        cj loop
next:
```

The remainder instruction uses both the floating-point and integer evaluation stacks, and cannot be overlapped with any integer processor instructions.

3.15 Debugging instructions

The T801 and more recent transputers include a set of instructions and a new processor flag to simplify the problem of setting breakpoints in transputer code (see Table 3.10). The problem arose from the fact that transputer instructions can be as short as 1 byte, and thus could not be replaced by a multi-byte jump to some

Table 3.10 Debugging instructions

Code	Mnemonic	Cycles	Description
0x00	jmp 0	3	jump 0, break not enabled
		11	break enabled, high priority
		13	break enabled, low priority
0xB1	break	9	break, high priority
		11	break, low priority
0xB2	clrj0break	1	clear *EnableJ0BreakFlag*
0xB3	setj0break	1	set *EnableJ0BreakFlag*
0xB4	testj0break	2	test if *EnableJ0BreakFlag*
0x7A	timerdisableh	1	disable high-priority timer interrupt
0x7B	timerdisablel	1	disable low-priority timer interrupt
0x7C	timerenableh	6	enable high-priority timer interrupt
0x7D	timerenablel	6	enable low-priority timer interrupt

debugging code. This is the procedure followed in conventional microprocessors, such as the 68000. As all the 1-byte instructions had already been allocated, INMOS had to give another meaning to an already existing instruction, and `j 0`, opcode `0x00` was chosen. This is very seldom used, as it has no effect on the flow of control, but does represent a potential descheduling point. An extra processor flag *EnableJ0BreakFlag* was added to control the interpretation of `j 0`. When this flag is not set, `j 0` has its usual interpretation. However, if *EnableJ0BreakFlag* is set, then execution of the `j 0` instruction results in a context switch. The values of *I* and *W* of the executing process are swapped with values taken from fixed locations in memory just above `MemStart`. The address map is shown in Figure 3.6.

A debugger can examine an executing applications program by loading the values of *I* and *W* of the breakpoint-handling process into low memory, and then replacing the appropriate instruction byte in the applications code by `0x00` and setting the *EnableJ0BreakFlag*. When the `j 0` instruction is executed, control will pass to the debugging process. The debugger may then return control to the application program by executing either a `break` or a `j 0` instruction.

The related instructions that disable the timer queues can be used to prevent timer events from interfering with debugging. If the queues are halted, no process waiting on the timers will be scheduled until the queues are enabled again. Processes

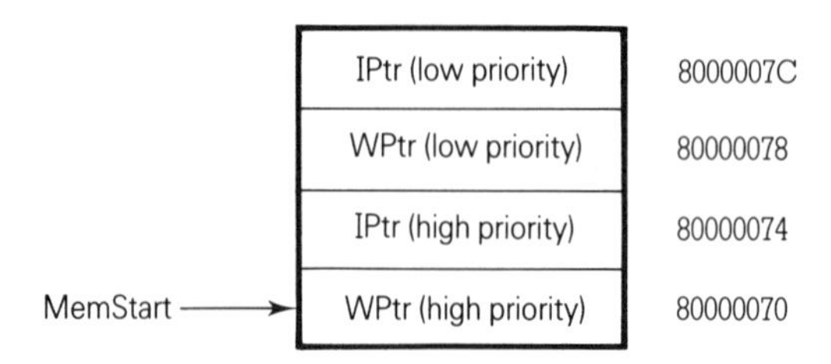

Figure 3.6 Breakpoint instruction memory locations

are not lost, but their execution may be delayed.

The ability to disable the timer queues has also proved useful in other contexts. The `move` instruction can execute for a long time when moving a large amount of data. Thus it has been made interruptible so that a low-priority `move` cannot unreasonably delay the running of a high-priority process. However, when the `move` instruction restarts it may repeat the copying of the word during which it was interrupted. Normally this is not a problem, but problems can occur if the instruction is being used to transfer data into a memory-mapped hardware fifo buffer, which can appear to receive extra data bytes. Halting the timer queues removes one potential source of interruption to the `move` instruction.

3.16 Processor identification

We have seen that many instructions behave identically on different processors of the transputer family. However, it would be very useful to be able to determine the type of processor on which code was executing, and whether or not it had a floating-point unit. The more recent transputer processors have the `lddevid`, load device identity, instruction to fulfil this need.

The `lddevid` instruction loads the A register with a small integer whose value depends on the processor type. The value loaded will be 0–9 on the T425, 10–19 on the T805, 20–29 on the T801, and 40–49 on the T225. On the T212, M212, T222 and T414 the instruction has no effect, on the T800 it sets B equal to C and leaves A undefined. We can write a simple instruction sequence that will identify the processor. First, initialize the evaluation stack so that $A = -3$, $B = -2$ and $C = -1$, then execute the `lddevid` instruction.

```
ldc -1
ldc -2
ldc -3
lddevid
```

We can then identify the processor from the new contents of the evaluation stack. On recent processors the device identity will have been loaded into A, the value of A pushed down into B, and the value of B pushed into C. The stack will be:

A device ID, positive integer
B -3
C -2

On the T800 the value of A will be undefined; it might happen to be a positive integer in the range of one of the device identities, but the value of B will be the same as the old value of C. The evaluation stack will contain:

A undefined
B 1
C -1

On the T212, M212, T222 and T414 processors the evaluation stack will not be altered and will contain:

A -3
B -2
C -1

The key to which group our processor belongs is in register B. If this contains -3 then the `lddevid` instruction has executed successfully and the processor ID is in A. If B contains -1 then the processor is a T800, if -2 then the processor is one of the old group on which the `lddevid` instruction has no effect.

Tables 3.11 and 3.12 show the instructions supported by the current range of 16-bit and 32-bit processors.

Table 3.11 Which processor has which instructions: 16-bit

	T212	T222	T225	M212
Extended Table 3.8	n	y	y	n
Debugging Table 3.10	n	n	y	n
FP support Table 3.9	n	n	n	n
Hardware FPU instructions	n	n	n	n
`fmul` instruction	n	n	n	n

Notes: The instruction set of the M212 is identical with that of the T222.

Table 3.12 Which processor has which instructions: 32-bit

	T400	T414	T425	T800	T801	T805
Extended Table 3.8	y	n	y	y	y	y
Debugging Table 3.10	y	n	y	n	y	y
FP support Table 3.9	y	y	y	n	n	n
Hardware FPU instructions	n	n	n	y	y	y
`fmul` instruction	y	y	y	y	y	y

Chapter 4

Low-Level Programming

This chapter describes some of the issues involved with programming a transputer at the lowest level. For many readers, the detailed operations described here will be performed by a software development environment. Other information will only be of essential interest to those writing a compiler or linker for the device.

4.1 Linking code

Code for the transputer may be produced from an assembler, or more likely from a high-level language compiler. In either case it is often necessary to combine the program with library routines, either provided within the software system or written locally.

The normal technique used to handle this is through the use of external variables. In a C program we may wish to refer to the routine `printf`. This is specified as being an external value, and until the program is linked with the C runtime system, the actual location of the routine `printf` is unknown. The linker satisfies the reference and patches the code to refer to the correct place.

In other microprocessors this is reasonably simple to achieve. The compiler can detect the use of an external reference, and simply generates a `JSR` or similar instruction followed by four empty bytes. This hole in the code is where the linker will patch the eventual location of the routine `printf`. Four bytes are sufficient for a 32-bit processor; although some processors have 16-bit variants of the `JSR` instruction these are rarely used by compilers as this would restrict the resulting program to be no more than 64 Kbytes long. The mechanism is slightly wasteful of space on many computer systems because the most significant byte following the `JSR` will always be zero unless more than 24 Mbytes of memory are being used. This is normally accepted as there are very rarely any 24-bit variants of `JSR`.

The situation on the transputer is rather different. An external routine will either be called via the `call` instruction, where the offset to `call` is part of the instruction itself; or else it will be used via the `gcall` instruction where the offset is loaded as a constant using `ldc`.

In the last chapter we discussed the variable length representation of numbers, and the way in which `pfix` is used to build up a number larger than a nibble. This can cause a difficulty where externals are to be used, as the amount of space to be left for the value of the external is unknown. In order to allow for arbitrary sized programs, 7 bytes would need to be left for each external, thus allowing the largest possible number to be patched in by the linker. Alternatively some limit of, say, 5 bytes must be chosen, thus creating a limit on the maximum size of program.

This problem also appears within the implementation of an assembler or compiler. Consider loading an address into A. The `ldc` instruction must be used to load the offset of a label, which is then converted to an address by `ldpi`. The assembler must know the offset of the label from the current program counter, and generate the correct value. This is easy in the simple example below, where the address of a string is being loaded:

```
        ldc L1
        ldpi
L1:     .byte "Hello World"
```

In this case the assembler knows the length of `ldpi`, and hence knows the value to be assigned for the offset of `L1`. But now consider a slightly more complicated version as follows:

```
        ldc L1
        ldpi
        ldc L2
        ldpi
L1:     .byte "Hello World"
L2:     .byte "Again"
```

The problem here is that the assembler does not know the offset of `L1` because there is a further `ldc` between it and the label. The length of this second `ldc` is not yet known, because the offset of `L2` is not known yet.

This particular example may be resolved by working out the offset of `L2` and then handling the offset for `L1`, but this cannot be done in all cases. Consider the slightly more general problem as follows:

```
L2:     .byte "Again"
        ldc L1
        ldpi
        call _printf
        ldc L2
        ldpi
L1:     .byte "Hello World"
```

Here the offset of `L1` depends on the offset used in the `call` instruction, which may not be known until link time, and on the size of the second `ldc` instruction. This in turn depends on the size of the first `ldc` instruction, which of course depends on the offset of `L1`. This type of circular dependency is the general problem which must be resolved.

The solution is reasonably simple but time consuming. A data structure representing the entire program is built in memory. Fixed length sections of code can be held as binary, but any label must be kept as a pointer to the label and an associated size. Initially all offsets are assumed to fit in one nibble, with no prefixes needed. A pass over the program is made, altering all those that require a larger offset to a suitable value. A further pass is then made, expanding those instructions that do not now fit because the previous pass expanded instructions. This process continues until no more changes need to be made.

This algorithm is the only one that is guaranteed to converge. The alternative, which entails shrinking code, has the advantage that iterations can be terminated at any point and you are left with a working, but non-optimal sized program. Unfortunately the algorithm can become caught and fail to converge.

This process can take a long time. A program of about 560 Kbytes needs about 300 passes before it converges. The process can be speeded up considerably by assuming an initial size of 3 bytes, and only expanding those instructions that are longer than this. However, the full optimization process is worthwhile for a finished program, as it removes about 10 percent of the program code; this 10 percent normally consists of wasted `pfix` instructions within crucial inner loops.

It is also apparent from this discussion that it is useful to perform this code size optimization within the linker, so that external variables do not lead to less than optimal programs. It is not vital that this is done, as externals are used less than other labels, but it does save some wasted code. In our experience it has been convenient to combine the function of assembler and linker into a single program.

4.2 Code generation issues

There are a number of differences between a transputer and other microprocessors that make life difficult for those charged with implementing high-level languages.

The biggest problem in designing a code generator for the transputer is the absence of any static data area. Normally a language requires two types of variable, commonly called local data and static data. In a language such as C, local variables are those declared within procedures and held on the stack; static variables are either called global or static in C and they persist beyond the procedure call.

The W register acts as a stack pointer, and hence local variables are stored at offsets from W. The `ldl` and `stl` instructions are used to refer to these variables, and the only design decision is whether to alter W whenever new variables come into scope, or whether to alter W once for each procedure and thus possibly waste stack space for the sake of less code.

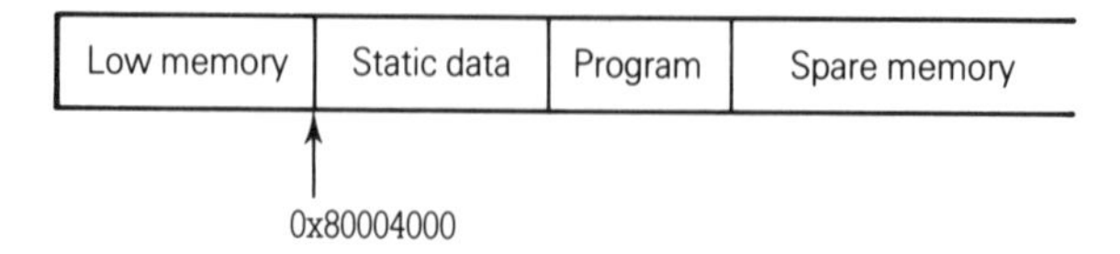

Figure 4.1 Allocation of static data area below program code at `0x80004000`

Accessing the static data area is more difficult. One solution is to insert references in the code to specific areas of memory. The actual memory address is determined by a linker which patches the code to refer to absolute memory. A loader ensures that these fixed memory locations are allocated and initialized correctly. Given this model, static data might be allocated starting at address `0x80004000`, as shown in Figure 4.1. Thus a code fragment to access a static value held at offset `0x10` from the static base would be as follows:

```
ldc  0x80004010
ldnl 0
```

This mechanism has the slight problem that the code for loading a large constant such as this is rather long, although sometimes better code can be generated by using `mint` followed by an `adc`. However, the main problem with this mechanism is that only one program can be loaded at a time, as the linker has the static base address built into it.

The standard solution for static data has been to use relocatable code. In this case the loader patches the code to refer to the correct offset and also adds relocation information to the code module. When the loader loads the program it allocates memory for the code and also for the static data. As it loads the program it uses the relocation information to patch the code once more to ensure that the absolute references to static data refer to the correct place in memory (Figure 4.2). There are also problems with this mechanism. First, the compiler must leave the maximum sized slots in the code for the loader to patch, as the value of the static pointer will not be known until load time.

Secondly, the code produced is not position-independent. Once the code has been loaded it cannot be moved. It is sometimes useful to be able to shuffle code segments in memory to make more space and this requires position-independent code. More importantly, it is often very useful to be able to download a copy of a program down a link into another transputer. In order to be able to do this using relocatable code the relocation information would need to be stored and downloaded as well.

Low memory | ... | Program code | Static data | ...

Figure 4.2 Allocation of static data area above program code

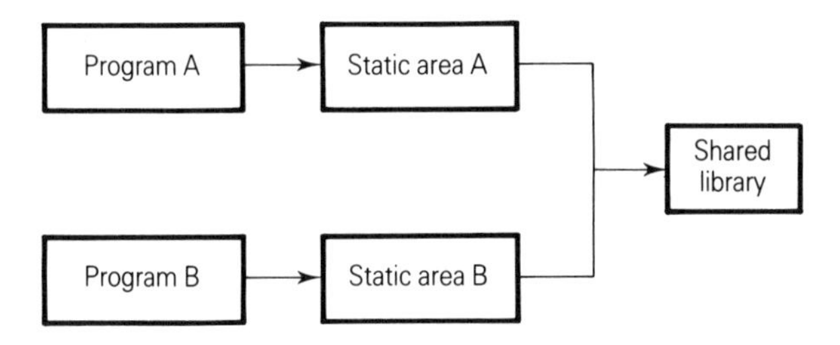

Figure 4.3 Allocation of static data areas and shared library for two programs

The final problem with relocatable code is that it is not possible to share it. If a transputer is running two programs, both written in C, they will both need to use the C runtime library. In the model described above each program will have a copy of the library and will relocate references to routines such as `printf` to the two different copies, even though the code will be the same.

One solution to this problem is to pass a static data pointer to every routine called. This means that for each routine the first parameter is always the static data pointer. An access to a static data item offset by `p` words from the static data pointer will require code of the form:

```
ldl n    -  load static data pointer
ldnl p   -  load data offset by p words from static
            data pointer
```

This has the disadvantage that an extra parameter must be passed to all routines. The advantages are that the code to access static data is reasonably compact, and it is position-independent. If the values of the static data pointer for two programs are the same, then they will share both code and data.

Although it is nearly always useful to share code, it is not usually useful to share data. A possible adaptation to this scheme is to arrange that the static data pointer points to an area of memory private to each program. Within this region are allocated static variables, while shared libraries are represented by pointers to the actual code (Figure 4.3). Two programs may have different static data pointers, pointing to two static data areas. Within each data area are pointers to the same shared code segments.

There is one further code generation issue concerned with making most efficient use of the stack. You will recall that there are two instructions `ldl` and `stl` used to access values from the stack, and that these are single byte instructions for stack offsets 0–15. Stack offsets in the range 16–255 require a single `pfix` instruction, making 2 bytes in total. Larger stack offsets require more `pfix` instructions.

It is clearly best to use the small offsets as much as possible, both to reduce the code size and to speed up execution. Consider the following C program fragment:

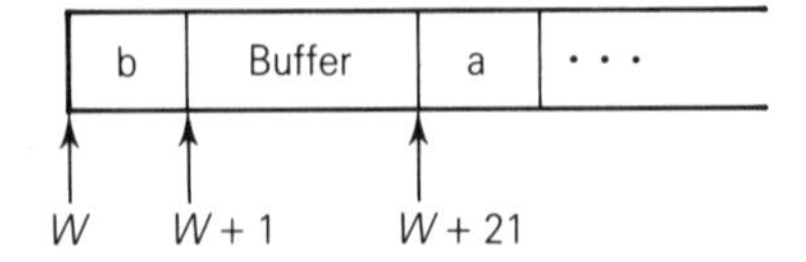

Figure 4.4 Local data access

```
foo()
{
        int a;
        char buffer[80];
        int b;
}
```

At first thought a compiler might allocate stack as in Figure 4.4. In this example, we need to use a `pfix` and then a `ldl` to access the value of `a`, which we will almost certainly need more often than every location in the array buffer. There are two major ways to achieve better code. The first is simply to sort the variables so that those used more often are closer to the bottom of the stack, and hence need fewer (or no) `pfix` instructions. This is often a useful thing to do, but it makes debugging a failed program that much more difficult because only the compiler knows where it has put things.

The second method is to use two distinct stacks for scalar values and for vectors such as C arrays and structures. In this case the stack might look like Figure 4.5.

Here we have kept a vector stack pointer *VS* as well as the normal stack pointer *W*. The *VS* pointer is used to implement a rising stack based at the end of the scalar stack. This arrangement is convenient because increasing the stack size increases the space available for both scalar and vector stacks, and because *VS* can also then be used to check for stack overflow conditions.

The disadvantage with using a vector stack is that the value of *VS* must be maintained somewhere, and access to vectors is more expensive than keeping them on the *W* stack. The value of *VS* may be held as a static variable, in which case it must be incremented when a routine is entered and decremented when it is left. Alternatively the value of *VS* can be passed up the stack, being suitably incremented where vectors are required. This approach has the advantage that when a routine exits, the *VS* is automatically restored to its previous value.

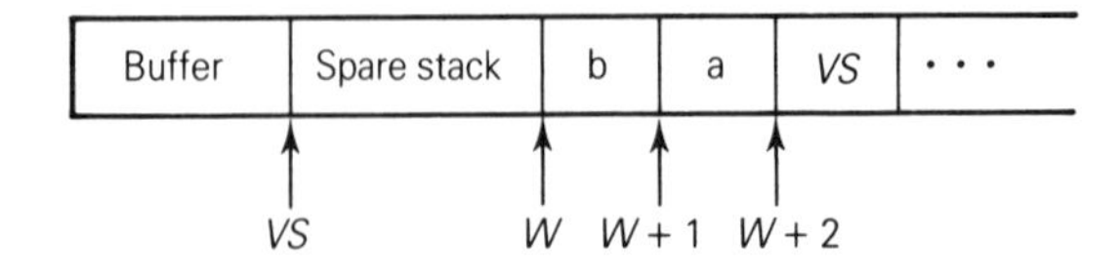

Figure 4.5 Local data access with vector stack pointer

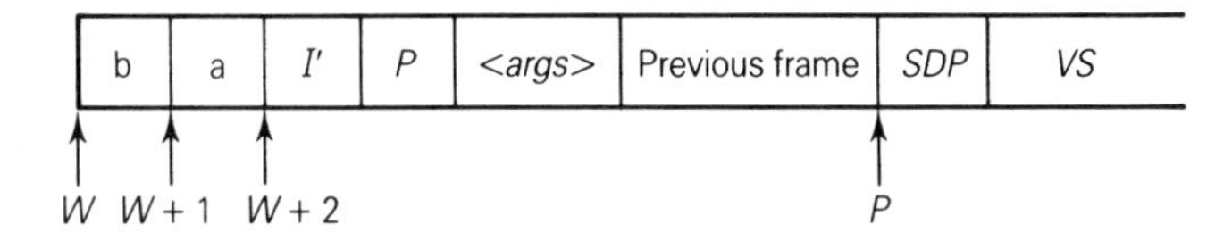

Figure 4.6 Stack frame when called routine does not use any vector stack space

A very useful code generation technique involves combining the static data pointer *SDP* and the vector stack pointer *VS* which are then both passed up the stack during each routine call. The compiler passes a pointer to a two-word block of memory up the stack which contains both *SDP* and *VS*. In the common case where *VS* need not be altered, this pointer is passed right through each routine. If some vector stack is required, a new two-word block is allocated on the stack and initialized with *SDP* and the altered value of *VS*. When the routine exits, the old block containing *SDP* and the original *VS* is used.

Thus, in a routine that did not use any vectors the stack would look as in Figure 4.6. The value P is passed as the first argument to the routine, and is stored by the call instruction just below the saved program counter I'. P itself refers to a two-word memory block in a previous stack frame which contains *SDP* and the current *VS*.

The stack for a routine that did use some vector stack space would look like Figure 4.7. Here the value of P passed by the previous routine is used only during the construction of the new two-word parameter block; *SDP* is copied directly and the new *VS* created by adding the size of the extra vector stack required to the old *VS*.

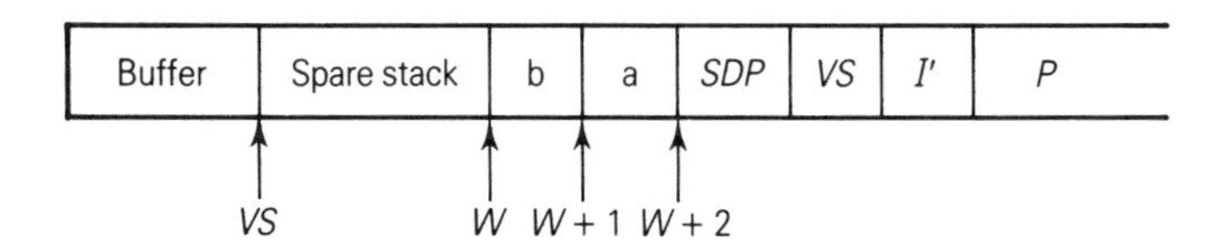

Figure 4.7 Stack frame when called routine uses vector stack space

4.3 Bootstrapping

When a transputer is reset it behaves in one of two ways, depending on the value of a pin prosaically entitled `BootFromROM`. If this pin is held high, then the transputer assumes that there is some ROM at the far end of its memory space and attempts to execute the instruction held in the top 2 bytes of the memory. If the pin is held low, then the transputer is to be booted from its links.

While a transputer is waiting to be booted from its links it behaves in a useful fashion familiar to those readers who have ever programmed in BASIC. If any link receives a byte of value 1, then the transputer awaits a further word. This is assumed to be a memory address at which the transputer then 'peeks'; the link then outputs a word which represents the contents of that memory location. In the same way a header byte of zero is used to indicate a 'poke' message; a word of address is sent down the link followed by the word of data which is then stored there.

If a value greater than one is sent as that first header byte, then the transputer reads that many bytes into its internal RAM. It then starts executing the code so loaded.

No matter how the initial bootstrap program is loaded into the transputer, once it starts running there are various operations it has to perform. The eventual aim of the bootstrap is either to start some ROM application or to load a larger program into the processor, either from the link that sent the bootstrap program or from a fixed link. Before it can do this various parts of the transputer must be initialized.

The first stage of initialization involves setting the scheduling queues to consistent values; special instructions are provided to set the high- and low-priority pointers and the timer queues. Then both the timers must be started and the *ErrorFlag* and *HaltOnErrorFlag* cleared.

The transputer comes up in this state irrespective of whether the chip was being simply reset (by pulsing the `Reset pin`), or whether it was being analyzed (by holding `Analyse` while pulsing `Reset`). However, the T805 and T425 transputers will clear *ErrorFlag* and *HaltOnError* if the transputer is reset without `Analyse` being asserted. Whether or not `Analyse` was asserted can be determined by the bootstrap process executing the `testpranal` instruction.

After a normal reset the bootstrap should initialize the bottom nine words of memory corresponding to the four links and the event channel. It should then continue to perform any external hardware operations required, such as writing to all memory locations to enable parity error detection. It should then jump to the ROM application or input the rest of the program from one of the links. An example bootstrap is given below.

```
testerr        clear ErrorFlag
clrhalterr     clear HaltOnError
mint           load value
sthf           initialize the high-priority queue
mint           load value
stlf           initialize the low-priority queue
mint           load value
mint           load address
stnl 9         initialize timer queue 0
mint           load value
mint           load address
```

```
        stnl 10            initialize timer queue 1
        ldc 0
        sttimer            zero timer
Prepare to initialize channels
        ajw 2              provide two extra words of workspace
        ldc 0              load zero
        stl 0              initialize loop variable
        ldc 9              load 9
        stl 1              loop limit
Loop to initialize channels
loop1:  ldl 0              get loop count
        mint               load address
        wsub               point to link word
        resetch            reset the channel
        ldlp 0             point to loop control block
        ldc loop2-loop1    load loop extent
        lend
loop2:                     loop
```

4.4 Debugging

As described earlier, the transputer may be analyzed rather than reset. This is a mechanism introduced by the designers of the transputer to enable a programmer to find out what has happened once a transputer has crashed. This provides some useful information which can be used by post-mortem debuggers.

If the processor was being analyzed, then the bootstrap would normally execute the instructions `saveh` and `savel`. These save the current values of the high and low scheduling queues' front and back pointers for later inspection. In addition some of the values that the registers had when the analyze occurred are made available to the bootstrap when it starts; the *A* register holds the old instruction pointer *I* and *B* holds the old workspace pointer *W*. Various internal lists are not reset and can be examined, as can the previous values of *ErrorFlag* and *HaltOnErrorFlag*.

The low end of memory contains some fixed locations which may be used by a post-mortem debugger. It is often useful to inspect some of these low memory locations after a transputer has crashed. In particular a debugger should look at the locations used to save the current state of a low-priority process when it is interrupted by a high-priority one. This will represent the state of the system when the last high-priority process ran, and is particularly useful when the program has jumped somewhere at random, leaving the previous *I* and *W* pointers invalid.

The low memory of a 32-bit transputer has the following structure:

Address	Description
`0x80000000`	link 0 output
`0x80000004`	link 1 output
`0x80000008`	link 2 output
`0x8000000C`	link 3 output
`0x80000010`	link 0 input
`0x80000014`	link 1 input
`0x80000018`	link 2 input
`0x8000001C`	link 3 input
`0x80000020`	event input
`0x80000024`	high-priority timer queue
`0x80000028`	low-priority timer queue
`0x8000002C`	W save area
`0x80000030`	I save area
`0x80000034`	A save area
`0x80000038`	B save area
`0x8000003C`	C save area
`0x80000040`	status save area
`0x80000044`	E save area
`0x80000048`	start of free memory (T414)

At this point, on the T414, memory may be used by the user. On the T800 the following extra locations are used to store information when performing a block copy:

Address	Description
`0x80000048`	two-dimensional block move save area
`0x80000070`	start of free memory (T800)

In general the information provided after a transputer crash is often of little use. In a language such as C it is very easy to reference non-existent memory accidentally, or to jump to data rather than code. In the first case some value is returned, which is only seen to be garbage much later in the program. In fact it is often highly beneficial to have some external control registers, or video display RAM, decoded to appear at location `0x00000000`. This at least gives some indication that a program is accessing through a null pointer, which is a common error.

In the second case the transputer just performs something, rather than providing some sort of 'illegal instruction' error. There is little that can be done in this case.

Although a post-mortem debugger is often useful, programmers often need to have a more complex debugger, showing what led to a problem rather than the aftermath. The hardware of the early transputers provides very little help here, and the best mechanism is either a full interactive debugger or some sort of trace vector.

A trace vector is an area of memory reserved for error diagnostics. A call to a `trace()` routine will store the current value of the instruction pointer *I*, along with any parameters passed to the routine, into the trace vector. The post-mortem debugger can then interrogate this trace vector to display the history of the last few calls made to `trace()` before the system crashed.

The trace vector mechanism is useful in parallel systems, where it is often difficult to arrange for even a simple print statement to arrive at the processor connected to a terminal before the system crashes. In many cases the very action of performing a print perturbs the system so that the bug no longer occurs, or manifests itself in some other fashion.

A fully interactive debugger is difficult to implement on the standard transputers, as it is not possible to insert a breakpoint instruction. A debugger will normally insert a breakpoint into a section of code by replacing the original code with some call to the debugger. When the breakpoint is to be passed, the original code is either interpreted or the original code replaced and executed.

The problem with this approach is that the breakpoint value to be inserted must be as small as the shortest instruction. Consider the case when the code contains a jump around the original instruction. If the breakpoint instruction were longer than the instruction it replaces, then the jump would encounter half of the breakpoint instruction. Consider setting a breakpoint on the `adc` instruction of the following code fragment:

```
        cj   loop1
        adc  4
loop1:  stl  5
```

Unless the breakpoint instruction were a single byte, the same size as the `adc 4`, then the `stl 5` would be corrupted. If the conditional jump were taken, then this corrupted instruction would be executed.

Normally, the only way in which a debugger can be called is by inserting a jump or call into the code, both of which are longer than a single byte and hence prone to the problem. This problem has been addressed in the more recent versions of the transputer, where a single byte instruction has been taken to implement a breakpoint instruction. The code chosen was previously that used for `j 0`, which is of course meaningless.

This is of no use in debugging the earlier transputers. Here two alternatives exist. The first is to ensure that sufficient gaps are left in the code, possibly before those code segments that correspond to a line of original source, and only to place breakpoint jumps in there. This is useful, but makes it difficult to step through code as the debugger must be able to analyze the generated code to determine the correct places where the breakpoint jumps may be inserted.

The second alternative is to cause a compiler to generate code which calls a debugging library. This library either returns immediately, or calls the debugger. This mechanism has two advantages. First, the library code may be shared by several processes. Each process may opt to enable the debug library and hence call the debugger. Secondly, it is possible to debug a process in another processor by ensuring that the debugging library sends messages via links from the test processor to the debugging one. However, the disadvantage of this approach is that the program is slowed down by the inclusion of the calls to the debugger, even when the debugging library causes an immediate return to the problem program. Additionally, the program must be compiled with the debug calls inserted, thus changing the nature of the code and possibly causing the problem to go away.

4.5 A disassembler

It was mentioned earlier that after a transputer has been reset the links may be used to peek and poke memory. Given this mechanism, it is often useful to construct a disassembler so that the original code can be viewed. The symmetric nature of the transputer instruction set makes this reasonably easy to achieve, and the following example provides the outline of a simple disassembler for the basic instruction set.

First we provide two arrays which contain the names of the transputer instructions. The first array covers the direct instructions corresponding to the nibble values 0–15. Although the names `pfix` and `nfix` are provided here they are not normally used as the disassembler decodes the numbers that these instructions are used to encode. Similarly the `opr` instruction is not normally printed because the disassembler will print the correct mnemonic for the extended instruction.

```
char *directfns[] = {   /* direct functions */
"j    ", "ldlp ", "pfix ", "ldnl ",
"ldc  ", "ldnlp", "nfix ", "ldl  ",
"adc  ", "call ", "cj   ", "ajw  ",
"eqc  ", "stl  ", "stnl ", "opr  "
};
```

In a similar fashion we provide another array which stores first the names of those instructions represented by a single byte. This byte always has the first nibble set to the code for `opr (0xF)` and hence these values represent the instructions identified by a single byte in the range `0xF0` to `0xFF`.

```
char *oper[] = {    /* one-byte operations */
"rev",   "lb",    "bsub", "endp",
"diff", "add",  "gcall","in",
"prod", "gt",    "wsub", "out",
"sub", "startp","outbyte", "outword",
```

The next set of instructions are much larger, and represent those that require 2 bytes. The first nibble of the first byte will always be the code for `pfix (0x2)`, while the first nibble of the second byte will always be the code for `opr (0xF)`. Thus these names match the sequence of bytes `0x21F0` to `0x21FF`, `0x22F0` to `0x22FF` and so on up to 0x2AFC. Note that not all codes are used and these slots are identified by a zero. These names are included in the same array since all these instructions are decoded as opr with an argument, where the argument may be a number formed by a pfix. The disassembler handles the pfix instruction before analyzing the opr.

```
     /* two-byte operations */
"seterr", 0, "resetch", "csub0",
0, "stopp", "ladd", "stlb",
"sthf", "norm", "ldiv", "ldpi",
"stlf", "xdble", "ldpri", "rem",
"ret", "lend", "ldtimer", 0,
0, 0, 0, 0,
0, "testerr", "testpranl", "tin",
"div", 0, "dist", "disc",
"diss", "lmul", "not", "xor",
"bcnt", "lshr", "lshl", "lsum",
"lsub", "runp", "xword", "sb",
"gajw", "savel", "saveh", "wcnt",
"shr", "shl", "mint", "alt",
"altwt", "altend", "and", "enbt",
"enbc", "enbs", "move", "or",
"csngl", "ccnt1", "talt", "ldiff",
"sthb", "taltwt", "sum", "mul",
"sttimer", "stoperr", "cword",  "clrhalterr",
"sethalterr", "testhalterr", "dup","move2dinit",
"move2dall", "move2dnonzero", "move2dzero", 0,
0, 0, 0, "unpacksn",
0, 0, 0, 0,
0, 0, 0, 0,
"postnormsn", "roundsn", 0, 0,
0, "ldinf", "fmul", "cflerr",
"crcword", "crcbyte", "bitcnt","bitrevword",
"bitrevnbits", 0, 0, 0,
0, 0, 0, 0,
0, "wsubdb", "fpldnldbi", "fpchkerr",
"fpstnldb", 0, "fpldnlsni", "fpadd",
"fpstnlsn", "fpsub", "fpldnldb", "fpmul",
"fpdiv", 0, "fpldnlsn", "fpremfirst",
"fpremstep","fpnan","fpordered","fpnotfinite",
"fpgt", "fpeq", "fpi32tor32", 0,
"fpi32tor64", 0, "fpb32tor64", 0,
"fptesterr","fprtoi32","fpstnli32","fpldzerosn",
"fpldzerodb", "fpint", 0, "fpdup",
"fprev", 0, "fpldnladddb", 0,
```

```
"fpldnlmuldb", 0, "fpldnladdsn", "fpentry",
"fpldnlmulsn", 0, 0, 0
};
```

Other floating-point instructions are executed by loading register A with a constant value and then calling `fpentry`; although this is normally represented by the sequence `ldc n; fpentry`, no attempt is made here to display the extended floating-point operation so identified.

The codes for `pfix`, `nfix` and `opr` are identified as these are handled specially.

```
#define f_pfix          0x2
#define f_nfix          0x6
#define f_opr           0xf
```

The following declarations are used to define space for the function part of an instruction, the instruction operand and a buffer where the instruction being handled is kept. The definition of a routine `gbyte()` is also found here. This routine is provided elsewhere and returns the next byte from the transputer memory. If this program were running on a transputer then it would merely return the byte held at location `curpos`.

```
int function;               /* function code */
int operand;                /* operand */
unsigned char ivec[8];
  /* buffer for decoded instructions */
int curpos;
  /* pointer to current memory location */
extern unsigned int gbyte()
  /* routine to return next byte */
```

The code proper starts with the procedure `decode()`. This is used to decode a set of bytes into correct function and operand parts. It handles any `pfix` or `nfix` instructions in the byte stream, calling `gbyte()` as many times as required. It also copies the bytes involved in the instruction into the buffer `ivec`.

```
void decode()
{
int i = 0;
int a_byte;
operand = 0;

for (;;) {
   a_byte = gbyte();
   curpos++;
   ivec[i++] = a_byte;
   function = a_byte>>4;
   operand = (operand << 4) | (a_byte & 0xf);
```

```
        switch ( (int) function ) {
           case f_nfix: operand = ~operand;
           case f_pfix: break;
           default    : return;
             }
        }
}
```

The final routine performs the disassembly. First the `decode()` routine is called to decompose the bytes. This also determines the length of the instruction. Most of the code here is to ensure that subsequent instructions are printed aligned. The current position in memory is printed, followed by up to 4 bytes of the instruction.

```
void disasm()
{
int loc = curpos;
int i, ilen;

decode();
ilen = curpos-loc;

printf("%8lx: ",loc);

for ( i = 0 ; i < min((int)ilen,(int)4) ; i++ )
   printf("%02x ",ivec[i]);
for ( ; i < 4 ; i++ ) printf("   ");
printf("         ");
```

The next part of the `disasm()` routine inspects the value of the function. If this identifies a direct instruction, then the name is printed, followed by the argument given by the value of `operand`. Otherwise if the function is `f_opr`, then the value of operand identifies the instruction and the name is printed from the second array.

```
if ( 0 <= function && function < f_opr )
  printf("%s %8lx    ",directfns[function],operand);
 elif (0 <= operand && operand <= 0xac && oper[operand] != 0)
  printf("%s",oper[operand]);
 else
  printf("UNKNOWN %2lx %8lx",function,operand);
printf("\n");
```

The final part of this example prints in hexadecimal any remaining byte from a particularly long sequence on the next line.

```
if( i < ilen ) {
   printf("         : ");
   for ( ; i < ilen ; i++ )
```

```
      printf("%02x ",ivec[i]);
    printf("\n");
  }
}
```

Chapter 5

Transputer Languages

The majority of transputers will probably end up in embedded computer systems, where the transputers act as the controlling processors for a device, such as a laser printer or a missile. These transputers will be completely under the control of the applications program; it is very unlikely that they will be used with any underlying operating system.

However, during the program development process it is necessary that some operating system facilities be available, such as access to a disk filing system and terminals, and that there should be facilities to run text editors, assemblers, high-level language compilers and debuggers. The simplest way to provide these is to use an existing machine as a host, running a server program that communicates with

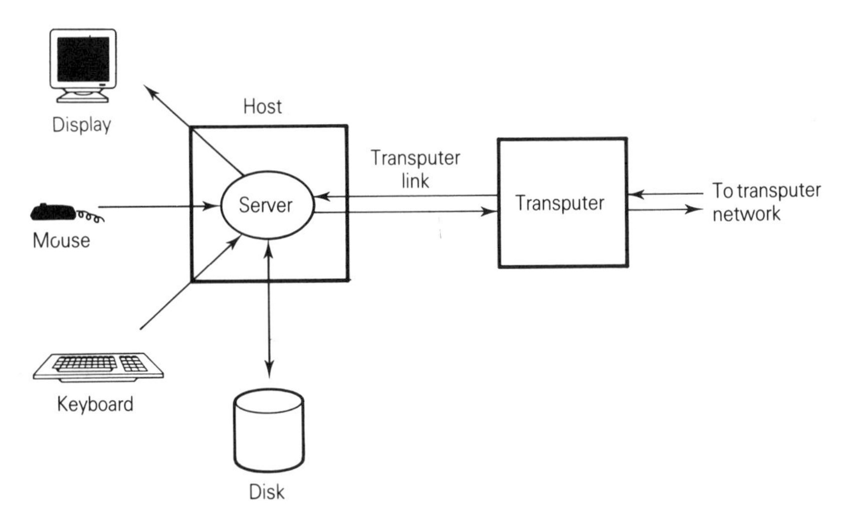

Figure 5.1 Host–server model: the host is connected to the transputer network by a single link

the transputer system. This also has the advantage that compilers and debuggers may be written to run on the transputer rather than being cross-compilers running on the host. Thus all that requires to be ported to a new host is a relatively simple server program. The server model also has the advantage that developers can continue to use the facilities of a familiar operating system, such as its command language and utility programs. Figure 5.1 shows a host–server model.

The first such system to be made available was the INMOS Transputer Development System or TDS. The TDS included a server, and an integrated occam programming environment including a compiler, an unusual 'folding' editor, a linker and a configurer. The server ran on a PC host, communicating with an INMOS B004 board consisting of a T414-15 with 2 Mbytes of RAM interfaced by a link adaptor to the PC bus. This product was provided with several versions of server programs in Intel 8088 assembler and in C.

A successor to the TDS server is a more general-purpose server called the 'iserver'. This is also available with a number of standalone tools such as compilers, editors, linkers and so on. The iserver will run on any computer system to which a transputer link can be interfaced. The server supports a protocol down the link which provides access to the screen, keyboard and files of the host. This protocol is used by the various tools that run on the transputer. Whenever the editor, for example, needs to read a file it calls a set of occam routines which send a request to the server. The server locates the file by calling the host operating system and sends it down the link to the editor. The editor then uses other occam routines which interact with the server to display characters on the host's screen and to read keyboard strokes.

The principal language provided with the TDS was occam, a high-level language that was designed for, and with, the transputer. Many INMOS documents describe it as the best possible language for programming the transputer, and for several years it was the only language available.

The name occam is derived from William of Occam (or Ockam, *c.* 1270–1349), an English scholar and philosopher. It was he who originated Occam's Razor, which states in its most familiar form, 'Entities are not to be multiplied without reason'. The maxim was never actually stated in this form by Occam, but he did say, 'It is vain to do with more what can be done with fewer'. An early variant, occam 1 or 'proto-occam', certainly lives up to this concept, although the latest version, occam 2, is much more complex.

Occam 2 provides most of the features that you would expect in a high-level language. However, many programmers used to languages such as Pascal or C will find occam unusual; there is no recursion, no structures or records except in input and output, no dynamic memory allocation, and no user-defined types. On the other hand, occam provides access to some of the transputer's facilities in a very clear and simple fashion. In particular it supports multiple concurrent processes, multiple processors, inter-process communications and the control of process scheduling.

Although INMOS has been a very strong advocate of occam, many programmers have preferred to continue using the more traditional languages, such as FORTRAN and C. This has been partially due to the difficulty that some programmers find in working with occam and the TDS, but must mainly be the result of the very small amount of existing occam software. For example, many large scientific packages are available only in FORTRAN, and important packages such as the X Window System are distributed only in C.

However, standard FORTRAN and C contain little if any support for parallel programming and this has had to be added, either by language extensions or in the runtime library support. In particular, a compiler for a standalone transputer system must support the server protocol, multi-tasking, internal and external channels, timers, and the distribution of code over a network of transputers. Two main ways of adding such features to conventional languages have found favor with software developers:

- Using the language unchanged, but adding parallel programming support through the runtime library.
- Altering the syntax of the language to add parallel programming constructs.

Below we describe occam in more detail and then go on to show how parallel features have been added to a conventional language, using C as an example. However, similar techniques have been used in FORTRAN and Pascal implementations.

5.1 Occam

As occam was designed, in parallel so to speak, with the transputer, the occam model of computation is based on communicating processes (Hoare, 1985). Processes have distinct memory spaces, but a mild degree of memory sharing is allowed in that two or more concurrent processes may read the same area of memory. However, if any process can write to memory, then no concurrent process is allowed to read that memory. Messages are passed between processes along channels, and messages act to synchronize processes. A process sending a message, or a process attempting to receive a message, will wait until the transaction is complete.

As well as explicit support for concurrent programming on a single transputer, occam includes a configuration language that allows the user to compile and configure a system that will run on a network of transputers. In this section we describe the main features of occam 2 in outline. This is not intended to be a complete or tutorial description of the language. Several excellent books on occam 2 now exist, and we particularly recommend Jones and Goldsmith (1988).

5.1.1 Processes

An occam program consists of one or more concurrently executing processes. Each process performs a sequence of actions, and then terminates. Each action may be an assignment, an input or an output. An assignment changes the value of a variable, an input receives a value from a channel, an output sends a value to a channel.

If no input is available from the channel, the process will wait. Similarly, if the process at the other end of the channel is unable to receive the output, the sending process will wait. Thus input and output provide both data transfer between processes and synchronization of processes. Both the sending and receiving process must be ready before the data transfer will take place.

Assignment process is shown in a familiar way by the := operator:

```
variable := expression
```

Input is designated by the ? operator:

```
channel ? variable
```

and output by the ! operator:

```
channel ! expression
```

There are two special processes. The SKIP process starts to execute, does nothing and then terminates. Thus it is equivalent to a NO-OP in assembly language programming. The STOP process starts to execute, does nothing, but never terminates. We will see how these special processes are used in the following sections.

5.1.2 Process sequences

Most conventional languages assume that statements, the 'lines of code', will be executed one after another in sequence. However, in occam this is not necessarily so, and the sequential nature of a process must be shown explicitly by the SEQ construct: Thus in occam we might say:

```
INT a :
CHAN OF INT chan1, chan1 :
SEQ
  chan1 ? a
  a := a * 5
  chan2 ! a
```

This program fragment declares a to be a variable, of type INT, and chan1 and chan2 to be channels, also of type INT. It consists of a sequence of three processes that input a value into a (from some other process), multiply it by the constant 5, and output it to another process. The declarations have the scope of the immediately following construct, and are indented to the same depth. Indentation by two spaces is used to show the scope of the sequential construct; in all cases occam uses indentation to show structure, where other languages might use `begin...` `end` or {......}.

5.1.3 Parallel processes

Occam processes may also execute concurrently, or in parallel. This is denoted by the use of the `PAR` construct:

```
INT a,b :
CHAN OF INT chan1 :
PAR
  chan1 ! a
  chan1 ? b
```

has the effect of copying the value of `a` into `b` and is equivalent to the following assignment statement:

```
-- declarations
SEQ
  b := a
```

The lexical order of the processes with the `PAR` construct is not important, and the processes will be started in an arbitrary order. Thus the following code fragment:

```
-- declarations
PAR
  chan1 ? b
  chan1 ! a
```

is exactly equivalent.

The `PAR` construct is a process that will terminate when all of its component processes terminate. Thus a `SKIP` process may be added to the `PAR` construct with essentially no effect. However, adding a `STOP` process would mean that the `PAR` construct would never terminate.

Occam says nothing about which of the processes in the `PAR` construct will execute first, or which will get a larger share of processor time. When it is necessary to give one process priority over another, then the `PRI PAR` construct must be used as follows:

```
PRI PAR
  process_a
  process_b
```

This construct is limited to two processes. Whenever the first process can execute it will; the second process will only execute after the first has completed or is waiting for an input or output.

The occam 2 PRI PAR is limited to two components, which map directly on to the high- and low-priority processes of the underlying hardware. If we want one set of processes to execute at high priority and another set at low, then PAR constructs must be nested in the PRI PAR as follows:

```
PRI PAR
  PAR
    high_1
    high_2
  PAR
    low_1
    low_2
    low_3
```

In transputer implementations of occam 2, a high-priority process will not be time sliced but will execute until completion or waiting for input or output. Only when all high-priority processes are unable to execute will the low-priority processes get their time slices. Because of the limitations of this mechanism, PRI PAR can only be used at the outermost level of a program, not within any enclosing PAR constructs.

5.1.4 Alternatives

It is often important to choose between alternative actions, depending on the order in which inputs become available. For example, a simple channel multiplexor copies the data from two inputs onto a single output (Figure 5.2). In occam this can be written using the ALT construct as follows:

```
INT a :
CHAN OF INT chanin1, chanin2, chanout :
ALT
  chanin1 ? a
    chanout ! a
  chanin2 ? a
    chanout ! a
```

This could also be written:

```
-- declarations
SEQ
  ALT
    chanin1 ? a
```

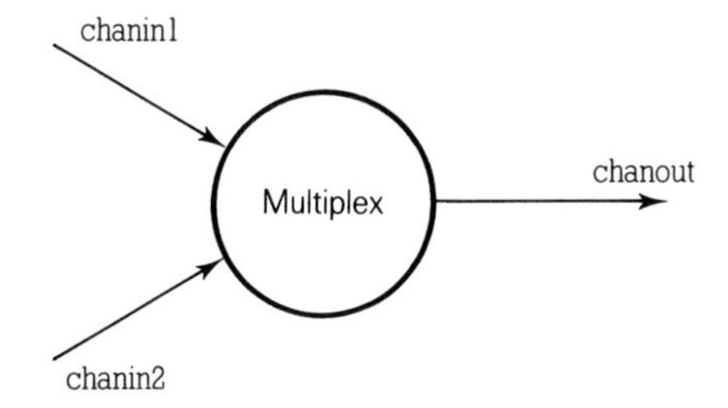

Figure 5.2 A two-input multiplexor

```
    chanin2 ? a
  chanout ! a
```

Whichever of the inputs becomes ready first will execute, and the appropriate process will read data into a. The whole process will then terminate, which is not what is required of a multiplexor. In order to keep the multiplexor running continuously, the above code must be embedded in a non-terminating loop:

```
-- declarations
WHILE TRUE
  SEQ
    ALT
      chanin1 ? a
      chanin2 ? a
    chanout ! a
```

Each of the alternatives may consist of a number of processes, in sequence, parallel, or indeed another ALT. Thus to tag each output with the input from which it came, we could write the following:

```
WHILE TRUE
  SEQ
    ALT
      chanin1 ? a
        SEQ
          chanout ! 1
          chanout ! a
      chanin2 ? a
        SEQ
          chanout ! 2
          chanout ! a
```

No relative priority is given to the processes within the ALT; the compiler can make an arbitrary decision as to which process will be executed if two or more become ready simultaneously. When priority is important the PRI ALT construct can be used. This gives priority to the lexically first element of the construct which will always be chosen for execution if it becomes ready before or at the same time as any other element.

```
PRI ALT
  chan1 ? a
    process.1
  chan2 ? b
    process.2
  chan3 ? c
    process.3
```

In the example above, if inputs are available simultaneously on chan1 and either of the other channels then process.1 will be started. Only if either chan2 or chan3 become ready without chan1 being ready will process.2 or process.3 be run. Unlike PRI PAR, PRI ALT is not limited to two components. However, all components of a PRI ALT except the first behave in the same manner as the components of an ALT. If the first component of an PRI ALT is chosen it is run at the priority of the enclosing construct, rather than at high priority.

5.1.5 Guards

In the ALT example above, the input statement that precedes each of the alternative actions is known as a guard. An ALT or PRI ALT is a sequence of guarded processes, where each guard is an input, a timer or a SKIP process with, optionally, a boolean condition. The guard:

```
go & chanin1 ? a
```

will be ready only if go is TRUE and chanin1 has an input.

A SKIP guard is always ready, and thus will execute whenever the boolean condition is true. The guard:

```
go & SKIP
```

is ready for execution whenever go is true. The boolean conditions within an ALT should not have the possibility of all being false at once. When the ALT statement has been entered the conditions cannot change, and thus the ALT process will act as a STOP if the boolean conditions on all the guards are false.

5.1.6 Arrays of processes

The FOR construct creates an array of processes, which can operate in sequence, parallel or as alternatives.

```
SEQ input = 1 FOR 8
  chan[input] ? buffer[input]
```

creates an array of eight sequentially executing processes that each input from a channel in the array `chan` into a corresponding cell of the array `buffer`. This is then equivalent to:

```
SEQ
  chan[1] ? buffer[1]
  chan[2] ? buffer[2]
    .     ?    .
    .     ?    .
  chan[8] ? buffer[8]
```

Thus the `SEQ...FOR` construct acts in a similar fashion to the `FOR` loop in Pascal or other high-level languages.

The `FOR` construct can also be used with `PAR` and `ALT`, so that:

```
PAR input = 1 FOR 8
  chan[input] ? buffer[input]
```

creates an array of eight concurrently executing processes, and:

```
ALT input = 1 FOR 8
  chan[input] ? buffer[input]
```

an array of alternatives, each waiting for input from one of the channels `chan[1]`... `chan[8]`. The loop limits in `PAR...FOR...` and `ALT...FOR...` must be constants, as occam 2 does not allow the dynamic creation of processes; the total number of processes must be known at compile time. This serious limitation is a feature of occam 2, rather than a consequence of the design of the transputer process scheduler.

5.1.7 Channel protocols

In occam 1 a channel was a stream of words or bytes, and the only outputs and inputs allowed were of these types. However, the channel itself did not have a type, and thus no type checking for channels was implemented. This led to confusion and error if what was put into a channel was not what was read from it.

The transputer implements internal channel input or output by copying data in memory. As we have discussed in Section 3.8, the second process to attempt input or output determines the number of bytes that are copied. If the processes do not agree on the number of bytes to be transferred, then the transfer will appear to take place but erroneous data will be copied. If we attempt the same erroneous communication over a hardware link it will fail in one of the transputers involved, as the link engine will be unable to complete the transfer.

Occam 2 provides a method for specifying the contents of messages sent along channels which can prevent this kind of error. An occam 2 channel is declared with a PROTOCOL as in the following examples:

```
CHAN OF BYTE a:
CHAN OF INT  b:
CHAN OF REAL64 data:
```

The above protocols declare that the channels will be used only for input or output of a single variable of the appropriate type.

A protocol can also be of array type:

```
CHAN OF [20]INT data :
```

which will receive or transmit streams of 20 INTs.

Where the size of the array to be transmitted is unknown, it is possible to declare a counted array protocol, which consists of an integer describing the size of the array, followed by that number of array elements:

```
CHAN OF INT::[]INT counted.chan :
[32] INT vector :
```

When data are transmitted the first word must be the count of the number of array items to be sent.

```
-- declarations
SEQ
  counted.chan ! 2 :: vector
```

will send a count of 2, followed by the first two elements of the array vector. Similarly, the count is the first data item to be received. Thus:

```
INT itemcount :
[20] INT datain :
SEQ
  counted.chan ? itemcount :: datain
```

will place the number of data elements received in the variable itemcount, and that number of INT elements in the array datain.

A protocol can also be a sequence of variables of the same or different types. This is achieved through the declaration of a sequential protocol, for example:

```
PROTOCOL DataPacket IS BYTE; REAL32; REAL32; INT:
```

In this protocol declaration the elements are separated by semicolons. The protocol thus declared can then be used in the declaration of a channel:

```
CHAN OF DataPacket InStream :
```

which can only be used for the input and output of streams of the sequence BYTE, REAL32, REAL32, INT.

5.1.8 Timers

A timer in occam is treated rather like a channel. Timers may be declared singly:

```
TIMER working :
```

or as arrays:

```
TIMER  [20]intervals :
```

Each timer can be read as if it were a channel returning a single integer value:

```
INT start, end:
SEQ
  working ? start
  intervals[2] ? end
```

This will read the value of the timer appropriate to the priority of the process in which the timer input occurs. However, when comparing times read in this way, it must be remembered that the number of ticks is read as an unsigned INT, with a number of bits equal to the word length of the transputer. Thus it is quite possible that the timer will roll over during any timing interval. Occam therefore provides an AFTER operator which causes a process to wait until the timer reaches a particular value:

```
working ? AFTER timeout
```

where timeout contains the value of the timer which will be waited for. Thus, to suspend execution for, say, 1000 ticks, one needs the following sequence:

```
TIMER s:
VAL INT wait IS 1000 :
INT now :
SEQ
  s ? now
  s ? AFTER now PLUS wait
```

where PLUS denotes unsigned addition. If the wait is required to be in seconds it must be computed using the number of ticks per second of that particular clock.

This form of wait can be used as a guard to provide for a timeout on some other input:

```
TIMER s :
CHAN OF INT inkey :
VAL INT timeout 2000 :
INT input, now :
SEQ
  s ? now
  ALT
```

```
        inkey ? input
          process.input(input)
        s ? AFTER now PLUS timeout
          write.string(terminal.screen,"Input failed")
```

This will accept an input from the channel `inkey`, provided that it occurs within `timeout` ticks. If not, the timer guard will become ready, and an error message will be displayed.

5.1.9 Placement

As occam is intended for the low-level programming of transputers, it contains features that allow the user to specify the position that variables occupy in the processor's memory, and also on which processor a particular process will execute. This is known as placement.

As we have discussed in Chapter 2, the hardware memory map of the transputer is byte-addressed, with signed addresses running from `MinInt` to `MaxInt`. The occam memory map is somewhat different. Addresses are unsigned words, running from 0 to the top of the address space. We can use the `PLACE` keyword to assign a memory address to a variable:

```
INT i:
PLACE i AT 28:
```

will place the integer variable `i` at hardware memory address `0x80000070` on a 32-bit processor, or `0x8038` on a 16-bit processor. This can be useful in allocating variables to the on-chip fast RAM, but tends to compromise the security of occam, as two processes can place variables at the same address, and can access them with no control.

Where this type of placement is most useful is in associating channels with hardware links. The link control words lie in the bottom eight words of the memory map, as in Table 5.1. For example:

```
CHAN OF INT chanin, chanout:
PLACE chanin AT 7:
PLACE chanout AT 3:
```

places the channels `chanin` and `chanout` on hardware link 3 input and output.

Placement of variables can also be used to gain access to memory-mapped peripherals; however, this is rather unsafe as more than one process can gain access to the peripheral in an uncontrolled manner. Also the occam standard gives no guarantee that a variable name appearing in an expression will only be accessed once. It is therefore much safer to treat memory-mapped devices as `PORT`s, which are extensions of the occam channel concept to locations in memory. If a `PORT` is declared and placed at an address, it is read and written to using channel input and output:

Table 5.1 Occam and hardware addresses of link control words

Control Word	Occam address	Hardware address	
		16-bit	32-bit
Link0Out	0	0x8000	0x80000000
Link1Out	1	0x8002	0x80000004
Link2Out	2	0x8004	0x80000008
Link3Out	3	0x8006	0x8000000C
Link0In	4	0x8008	0x80000010
Link1In	5	0x800A	0x80000014
Link2In	6	0x800C	0x80000018
Link3In	7	0x800E	0x8000001C

```
PORT OF INT memloc:
PLACE memloc at #10000000 :
INT i;
SEQ
  memloc ? i
```

This code fragment will read the contents of the specified memory location into `i`. It guarantees that only one read access will be made to `memloc`.

5.1.10 Configuration

Allocation of processes to processors is normally known as configuration. The occam configuration language uses an extension of the `PAR` construct, `PLACED PAR`. This can only be used at the outermost level of an occam program, and specifies that a process is placed on a particular processor. The code fragment:

```
-- declarations
PLACED PAR
  PROCESSOR 0
    task0
  PROCESSOR 1
    task1
```

declares that `task0` runs on `processor 0`, and `task1` on `processor 1`. The `FOR` construct may be used with `PLACED PAR`, and so to run the same process on ten transputers, `PROCESSOR 0-9`, we can write:

```
PLACED PAR i = 0 FOR 10
  PROCESSOR i
    task
```

Links can be declared outside a processor, and so can be used to tie processes together that execute on different processors. If the process `task` is defined with two parameters, an input and an output channel, then:

```
[11] CHAN OF INT32 chan :
PLACED PAR i = 0 FOR 10
  PROCESSOR i
    task(c[i],c[i+1])
```

declares a pipeline of `task` processes executing on different processors, and passing data from one to the next through the elements of the channel array `chan[]`. These channels must be made to correspond to physical links by placement. As an example the input link below is defined as `link 1`, and the output as `link 2`:

```
VAL INT Link1In IS 5:
VAL INT Link2Out IS 2:
[11] CHAN OF INT32 chan :
PLACED PAR i = 0 TO 10
  PROCESSOR i
    PLACE c[i] AT Link1In :
    PLACE c[i+1] AT Link2Out :
    task(c[i], c[i+1])
```

Here the constants `Link1In` and `Link2Out` have been defined as the occam addresses of the channel words corresponding to link 1 input and link 2 output.

5.2 Runtime library support for parallelism in C

As we have discussed earlier, it is possible to add support for parallel programming to a conventional language either by modifications to the language syntax, or by leaving the language unchanged and providing runtime library functions for process creation and inter-process communication. The language we use here to illustrate how C may be extended by adding runtime library functions is Parallel C, from 3L Ltd. This is a transputer C compiler that implements the language described in Kernighan and Ritchie (1978), not the later ANSI standard. The extensions to support parallel computation are provided in the runtime library, and in a configurer that will construct systems to run on transputer arrays. This is the compiler that forms the basis of the INMOS D511, 611 and 711 products. Helios C and Logical Systems C are similar to 3L C in their approach to providing parallelism.

A Parallel C application consists of a collection of one or more concurrently executing tasks. Each task has its own region of memory for code and data, a vector of input ports, and a vector of output ports. The tasks are complete C programs, linked together with all or part of the runtime library. How these tasks are distributed over an array of transputers will be discussed in the section on the configuration language.

Within each task there can be one or more concurrently executing threads. These are processes that are similar to the components of a `PAR` structure in occam, but able to share global data. In Parallel C many instances of the same function can be active simultaneously as threads, each with its own copy of local data. Threads can share global data and other global resources such as channels and the runtime library. Semaphores are provided to control access to these global resources; channel input or output and timer support are provided as runtime library routines. Each package of library routines has associated with it an 'include' file that defines various useful constants.

5.2.1 Creating threads

A thread is created in Parallel C by a call to the `thread_start` function:

```
thread_start(fn, ws, wssize, flags, nargs, arg1,
                ..., argn)
```

Here `fn` is the name of the function to be executed as a thread, `ws` is a pointer to a workspace, and `wssize` the size of that workspace in bytes. The `flags` argument contains either `THREAD_URGENT` or `THREAD_NOTURG`, to indicate if a high- or low-priority thread is to be started. A variable number of arguments may be passed to the function by `thread_start`, the number of arguments is given in `nargs`, and the arguments in `arg1` to `argn`. A simpler form of thread creation is provided by the `thread_create` function:

```
char *thread_create(fn, wssize, nargs, arg1,
                        ..., argn)
```

This routine starts a thread at the same priority as the calling routine, with workspace taken from the heap. The function returns a pointer to that workspace, or `NULL` if insufficient space is available and the thread cannot be started.

Threads stop when they return to the calling routine, or by a call to `thread_stop`. Threads stopped by `thread_stop` can be restarted by a call to `thread_restart(p)`, where `p` is a pointer to the workspace of the thread. Threads may also be suspended when it is necessary to reset a channel, and can be restarted by `thread_restart`. Table 5.2 summarizes the Parallel C thread routines.

Table 5.2 Parallel C thread routines

`thread_start`	start a general thread
`thread_create`	start a thread simply
`thread_stop`	stop a thread
`thread_restart`	restart a stopped thread
`thread_deschedule`	pause a thread for about one time slice
`thread_priority`	return the priority of the current thread

5.2.2 Inter-process communication

The procedures that provide channel input and output in Parallel C are shown in Table 5.3. Each procedure is passed an argument of type `CHAN`, a predeclared type which represents the address of a channel control word. Channel addresses can be obtained in three ways:

- The addresses of channels connected to the hardware links of transputers are contained in predefined constants `Link0Input...Link3Input` and `Link0Output ...Link3Output`.
- The addresses of channels associated with the vectors of input and output processes of a Parallel C program are passed to the program through the main program parameter line.
- Any integer variable may be used as a channel control word. Channels defined in this way must be initialized before use by a call to `chan_init`.

Table 5.3 Parallel C channel support

Input and output routines	
`chan_in_byte`	input a byte from a channel
`chan_in_word`	input a word (4 bytes)
`chan_in_message`	input a stream of bytes
`chan_out_byte`	output a byte to a channel
`chan_out_word`	output a word to a channel
`chan_out_message`	output a message to a channel
Timed input and output	
`chan_in_byte_t`	input a byte or timeout
`chan_in_word_t`	input a word or timeout
`chan_in_message_t`	input a stream of bytes or timeout
`chan_out_byte_t`	output a byte or timeout
`chan_out_word_t`	output a word or timeout
`chan_out_message_t`	output a message or timeout
Initialization	
`chan_init`	initialize a channel word
`chan_reset`	reset a channel

5.2.3 Semaphores

As stated above, Parallel C threads can share global resources, such as data, channels and the runtime library. Thus a mechanism has had to be implemented to resolve access conflicts. This is based on the semaphore, a global memory location that each thread may examine and alter in a single uninterruptible instruction. Each semaphore may have a queue of threads waiting on it. When the currently active thread has ceased to need the resource protected by the semaphore it signals this, and one of the waiting processes is selected for execution. Of course, one must be very careful that all threads needing a resource wait on the appropriate semaphore, and that all threads signal the semaphore on ceasing to need the resource.

Parallel C provides a predefined data type `SEMA`. A semaphore, `s`, declared as this type must be initialized by a call to `sema_init` as follows:

```
sema_init(&s,v)
```

This sets up the semaphore so that the queue of threads waiting on it is empty, and the value of the semaphore is `v`. When a thread requires a resource controlled by a semaphore it makes a call to `sema_wait(&s)`. If the value of the semaphore is zero, it is left unchanged and the thread added to the list of threads waiting on `s`. If it is non-zero, the value is decreased by 1, and the thread continues its execution. When the thread has completed its use of the resource it must call `sema_signal(&s)`. If there are other threads waiting on the semaphore its value will be zero. One of these threads will be reactivated, and the value of the semaphore left as zero. If there are no threads waiting on the semaphore its value, which may be zero, will be increased by one.

In most cases only one thread at a time may use a resource. The semaphore should be given an initial value of 1. Each thread either finds the semaphore count at 1, and continues its execution after decreasing it to 0, or finds it at 0, when it waits. When a thread calls the `sema_signal` routine it will always find a semaphore value of 0. The semaphore value will remain unchanged if there are other threads waiting, otherwise it will be increased to 1.

Table 5.4 Parallel C semaphore routines

`sema_init`	initialize a semaphore
`sema_wait`	wait on a semaphore
`sema_wait_n`	perform *n* wait operations
`sema_signal`	signal a semaphore
`sema_signal_n`	perform *n* signal operations

It is interesting at this point to see how semaphores could be implemented if this facility were not provided in the runtime library. The implementation shown below is based on that provided in the Helios operating system; it is not necessarily similar to the way that semaphores are actually implemented in Parallel C. The procedures developed below would have to be run at high priority, so that there is no possibility of corruption of the semaphore mechanism by simultaneous calls from two or more processes.

A semaphore consists of a count, and a (possibly empty) list of processes waiting on the semaphore. Thus a semaphore structure may be declared as follows:

```
typedef struct SEMA {
        word Count;  /* semaphore counter      */
        Id *Head;    /* head of process list   */
        Id *Tail;    /* tail of process list   */
           } SEMA;
```

The structure Id is used to construct the linked list of processes, and has the following form:

```
typedef struct Id {
   word state; /* save area for process state   */
   Id *next;   /* pointer to next item in list */
} Id;
```

To initialize the semaphore the count must be set to the required value (usually 1), and an empty process list set up:

```
void sema_init(SEMA *sem,  word count)
{
  sem->count = count;
  sem->Head = NULL;
  sem->Tail = (Id *)sem;
}
```

The cast (Id *) allows the tail pointer to refer back to the semaphore record when the linked process list is empty.

To wait on the semaphore the value of the count must first be decremented and tested. If the result is less than zero the process must be added to the queue waiting on the semaphore, otherwise the process may proceed:

```
void sema_wait(SEMA *sem)
{ Id w;
 sem->Count--;
 if (sem->count < 0)
   {
    w.next = Null;
    sem->Tail->next = &w;
    sem->Tail = &w;
    w.state = thread_stop();
   }
}
```

The function thread_stop() removes the current process from the scheduling queue, saving its *W* register and priority so that it can be restarted.

To signal the semaphore that the process has completed the use of the shared resource, the sema_signal() function must increment the count, and start the next process in the queue:

```
void sema_signalignal(SEMA *sem)
{
    sem->Count ++;
    if (sem->Count < 1)
       /* there are waiting processes */
       {
         Id *w = sem->Head;
         sem->Head = w->next;
         if (sem->Head == NULL)
               sem->Tail = (Id *) sem;
          thread_restart(w->state);
       }
}
```

Again, the casts are used to allow the storage of the address of the semaphore record in the tail pointer of the linked list. The `thread_restart()` function reschedules a process which has been descheduled by `thread_stop()`.

5.2.4 Alternation

Parallel C provides no direct implementation of the transputer's use of alternatives, or of the occam ALT construct. Thus, if an action is required when the first of a possible set of events occurs it is necessary to set up a thread waiting on each of the events. These threads are then controlled by the use of a semaphore so that only one is allowed to execute.

Let us take as a simple example the two-input multiplexor of Figure 5.2. In occam this can be implemented as a procedure:

```
PROC mult2 (CHAN OF INT chanin1, chanin2, chanout)
  INT a:
  WHILE TRUE
    ALT
      chanin1 ? a
        chanout ! a
      chanin2 ? a
        chanout ! a
:
```

The shared resource here is the output channel `chanout`, which is protected in occam by the fact that only one of the components of the ALT will execute each time round the loop. In C we must use a semaphore to protect the channel. The C multiplexor function can then be written as follows:

```
void mult2(chanin1, chanin2, chanout)
CHAN *chanin1, chanin2, chanout;
```

```
{
 SEMA sem, *s;
 char *test;

 s = &sem;
 sema_init(s, 1); /* initialize semaphore */

 test = thread_create(copy,10,2,chanin1,chanout);
   if (test == NULL) exit();

 test = thread_create(copy,10,2,chanin2,chanout);
   if (test == NULL) exit();

 for (;;) timer_delay(1000000000);
  /* loop that uses very little processor time */
}
```

This function starts by initializing the semaphore with a count of 1. It then creates two threads, each waiting on one of the input channels. The copy function reads the input from its channel and, if necessary, waits until the output channel is free:

```
void copy (s, chanin, chanout)
SEMA *s;
CHAN chanin, chanout;

{ int data;
 for (;;)   /* non-terminating loop */
  { data =  chan_in_word(chanin);
        /* wait for input */
    sema_wait(s);
        /* wait for output to be free */
    chan_out_word(data);
        /* write out the data */
    sema_signal(s);
        /*signal output channel free */
  }
}
```

5.2.5 Configuration

A Parallel C application is made up of a number of tasks, distributed over a number of processors, and connected together by unidirectional channels. The channels that connect processes on different processors must be associated with physical wires that can carry one channel in each direction. A host processor must be declared running the server task, here called `afserver` for historical reasons. Clearly, there must be a path through the 'wires' (actually the transputer links) to each processor, so that the system is connected and can be booted. The configuration instructions are processed by the 'config' program, which reads in the appropriate task images produced by the 3L compilers and linker, and produces a file that is suitable for booting a transputer array.

For example, if we wish to run a program `compute` on a single transputer, the configuration instructions required will be as follows:

```
processor host
processor root

wire ? host[0] root[0]

task afserver ins=1 outs=1
task compute  ins=1 outs=1

place afserver host
place computer root

connect ? afserver[0] compute[0]
connect ? compute[0] afserver[0]
```

The `processor` statements name the processors that are in the system. One processor must always be named `host`; it is assumed that it is of type `pc` and will run the `afserver` task. There may be other PCs connected into the system and they must be declared with `type=pc`, so that the system does not attempt to boot them as transputers. To declare two transputers, a host PC and another PC, we would write:

```
processor host
processor transputer1
processor transputer2
processor other_PC type = pc
```

The `wire` statement specifies the physical link connections between processors. A wire can be given a name, but this is often not necessary and a `?` can be substituted for the name. So, if in the example above each of the processors is connected to its neighbor in the list by one link, then we would write:

```
wire ? host[0] transputer1[0]
wire ? transputer1[1] transputer2[0]
wire ? transputer2[1] other_PC[0]
```

The order of the processors in each wire declaration is not important, as the physical link is bidirectional.

The `task` statement declares programs that are to be loaded on the transputer network, and can also be used to specify the amount of memory required by each task. This is essential when more than one task is to be loaded on a single transputer. The `task` statement may also specify that file from which the task code is to be loaded.

The `place` statement places an individual process on a particular processor, the `afserver` server task must always be placed on the host PC. A process may only be placed on one host; if we require several copies of the same code, then each copy must be given a different task name, but may be taken from the same file, for example:

```
task in1 file="inprocess.b4"
task in2 file="inprocess.b4"

place in1 transputer1
place in2 transputer2
```

`Connect` statements then specify the way in which processes will be connected by channels. If the processes are on different processors, then the physical links or 'wires' will be used. If the processes are on the same processor the connections will be made by internal channels. To make a bidirectional connection two `connect` statements are necessary. For example:

```
connect afserver[0] compute[0]
connect compute[0] afserver[0]
```

makes a bidirectional connection between the `afserver` task and the `compute` task. As these tasks are on the `host` and `root` processors respectively the connection will be made along the 'wire' connecting these processors. This has been defined above as joining link 0 on `host` to link 0 on `root`.

The configuration file is used as the input to the 3L configuration program which outputs a transputer network program containing a distributing loader. This program is loaded simply by copying it byte by byte into the transputer at the root of the network. The distributing loader then bootstraps each processor in the network, and then loads on it the appropriate code.

5.3 Adding parallel syntax to C

An alternative approach to making a parallel processing version of C is to extend the language syntax to support parallel processing structures. This has the disadvantage that the code produced is not standard C, which limits portability. On the other hand, the addition of parallel constructs can lead to a much more natural programming style than relying on runtime library extensions. The example we use here is the Parsys par.c compiler.

5.3.1 Concurrent execution

To execute C statements concurrently in par.c they are enclosed within a `par` construct, just as in occam. For example:

```
par {
      buffer();
      read_it();
      write_it();
    }
```

will run the routines `buffer()`, `read_it()` and `write_it()` concurrently.

5.3.2 Channel pseudo-variables

Channel input and output is handled by channel pseudo-variables. These can be used where variables are normally used, but will have the effect of reading or writing a channel. Pointers to channels are also allowed. A channel must be initialized by a call to `resetch`, and can then be used for input or output:

```
channel chan1;

main()0
{ resetch(&chan1);
  resetch(&chan1);
  par {
       process_1();
       process_2();
       }
}

process_1()
/* output a constant to a channel */
{
     chan1 = 256345;
}

process_2()
/* input an integer from a channel */
{ int i;
  i = chan1;
  printf("%d\n",i);
}
```

The data type that is sent over the channel depends on the type of the expression that is assigned to the pseudo-variable. Clearly, the sending and receiving process must agree on the type.

Pointers to channels are used when a channel has to be associated with a hardware link. Thus to set up a channel at the `link 0` input we can declare a pointer to it and initialize that pointer with the correct hardware address. Data can then be read from the link by using that address on the right-hand side of an assignment, so that the function:

```
int get_word()
{
  channel *link0in = (channel *) 0x80000010;
  return(*link0in);
}
```

inputs a word from hardware link 0.

5.3.3 Timers

The two transputer timers are accessed by a pseudo-variable called `timer`, which is read depending on the priority of the process. Three different operations are possible to set, read or wait on the timer:

```
int i;

(int) timer = 0; /* set timer to 0
                    the (int) cast is required */

i = timer; /* read the timer into i */

timer = 100; /* wait until timer == 100 */

timer += 100; /* sleep for 100 ticks */
```

5.3.4 Alternation

Alternation in par.c is implemented in a similar way to that of occam. An **alt** keyword introduces a group of statements, each consisting of a guard and a block of code. If one or more guards becomes ready, then the code belonging to the lexically first ready guard in the sequence will be executed. There are three types of guards: boolean, channel and timer.

A boolean guard consists of the keyword **guard** followed by a boolean expression and terminated by a colon:

```
guard a > b :
```

This type of guard will be ready if the boolean expression is true; the booleans are evaluated once when the `alt` statement is entered.

A channel or timer guard has an optional boolean part with a channel address or a timer expression:

```
guard &chan0 :
guard a > b,  &chan1 :
guard &chan2, complete == TRUE :
guard timer += 1000 :
```

We may now write our simple multiplexor function in this flavor of C:

```
int mult(chan1, chan2, chanout)
channel chan1, chan2, chanout);
{ int i;
 for {;;}
 alt{
     guard &chan1 :
      chanout = chan1;
      break;
     guard &chan2 :
      chanout = chan2;
      break;
     }
}
```

Chapter 6

Software Environments

The previous chapter described a number of different languages used to program parallel transputer-based computers. Although the languages were all different, they embodied a common view of the context in which the transputers were to be used. This was the host–server model, where a host computer runs a server program which communicates with the transputer network.

The program running within the transputer network can perform input/output (I/O) in one of two ways. First, it can request services from the server using a suitable protocol. The server then performs the actions needed on the program's behalf, and sends back the data so requested. This protocol can be used to access files, the keyboard, a mouse or any other peripheral connected to the host.

Secondly, a program can communicate directly with the hardware on which it is running; in this case I/O is normally performed by direct manipulation of hardware registers.

The host–server mechanism is highly efficient, as there is no overhead in memory space or execution time taken by any system software. Many functions normally provided by an operating system or other low-level system software are not required in the transputer because they are already provided in the instruction set. These include process scheduling, high- and low-priority processes and inter-process communication. It is reminiscent of the early stages of microcomputer technology where a program interacted directly with the hardware with no supporting layer of system software. Such a scheme is used very commonly today whenever a microprocessor is being used in any embedded situation. This embedded use is the target for any semiconductor manufacturing company – it is far better for them to be able to place a chip in every video cassette recorder than it is to have one in a best-selling personal computer, merely because the number of units sold is so large.

INMOS always had this embedded marketplace in mind when it designed the transputer, and the host–server model is a system designed to aid in creating stand-alone systems that do not need I/O, such as keyboards, screens and disk files. A good example of such an embedded application is a laser printer controller.

There are, however, many cases where the facilities provided by the naked transputer are insufficient. First, the programmer must be fully aware of the hardware on which the program is to run. The number of processors and the topology of the processor network is built into the program. The program cannot run without this exact hardware, and if more processors are added the program cannot use them until it is recompiled. Clearly this is of little importance in an embedded system, but becomes more important when transputers are used in generalized applications.

These problems are becoming exacerbated by the way in which computer system builders are employing the transputer. In some cases very large systems are being built to solve a particular problem. These systems can have as many as 1024 transputers and several gigabytes of RAM; such supercomputers are far too expensive to be used for just one carefully crafted problem. Instead, owners of such machines would like to be able to buy standard application packages that will work on the particular hardware available, be this one transputer or one thousand. Having purchased the application, they would like to be able to buy extra transputers and have the application run faster. This is not possible with an application that runs on a naked transputer.

A second problem arises from the lack of standards available when programming at such a low level. For example, it is possible to clear the screen and move the cursor around on a transputer connected to a personal computer by sending certain control codes to the screen. These same control sequences may well not work on a transputer connected to a workstation or to a large minicomputer. The way in which files are specified will be very different depending on the operating system running on the host computer; UNIX, MS-DOS and VMS all use different mechanisms. These filenames will need to be quoted explicitly within a transputer application. A program designed to run on four transputers connected in a certain topology and hosted on a VAX will therefore not run on exactly the same transputer hardware but connected to a personal computer without altering and recompiling.

The large computer systems mentioned above commonly have electronic link configuration. Such systems should be available to run a single large problem on all the transputers, and when that job completes it should be possible to reconfigure the system into four or five smaller networks and run different programs in each. Ideally this reconfiguration – as well as the management of the processor resource – should be handled by system software.

Another problem concerned with programming naked transputers comes from the application program being required to handle naked peripherals. At the lowest level this means that any program running in a transputer that has hardware capable of causing an event by toggling the `EventReq` input must contain code to cope with this event. At a higher level it means that application programmers must learn about the quickest way to provide reliable message passing between two processes running on different processors, especially where there is no direct link connection between them. At a higher level still it is easy for an application programmer to make poor use of a piece of hardware such as a disk drive by getting complex tuning values such as the interleave factor wrong.

In all these cases it is useful to have some system software contained within the transputer. Such a system might provide some or all of our requirements: an efficient, standardized interface, a way of running programs independent of the topology and adequate control of hardware resources. A full operating system will normally add to this a set of standard tools and one or more user interfaces.

6.1 Express

Some of the early development work in parallel computing took place at Caltech. Much of this work was initially based on Intel Hypercubes and *N*-cube machines. One of the early outcomes of this work was the requirement to be able to run parallel programs on these different machines but within the same fundamental operating environment. The system that evolved there has been ported to the transputer, and is marketed commercially as Express.

Express is an operating environment that runs on a number of parallel processors. It provides certain operations normally found within an operating system in order to provide the basic functionality required by parallel programs. These operations include the ability to communicate, to share data, read files, perform graphics and so on.

Although this may sound like the specification of an operating system, Express is not an operating system in the conventional sense. It does not provide mechanisms for handling terminals or disk units at a fundamental level; instead it passes such requests back to an operating system running on some host machine. This has the advantage that the user of Express need not learn a new operating system because all of the facilities of the familiar host operating system are still available. It does, of course, mean that the host must be connected to the transputer network in some way.

Express is best described as a set of tools and utilities designed for parallel processing. These are as follows:

- A set of both high- and low-level communications primitives used to send messages between processors.

- A transparent I/O system, allowing any transputer in the network access to the facilities of the host operating system, as if the program running on the transputer were actually running on the host.

- A parallel graphics system available to all the transputers on the network. Both low-level graphics primitives and high-level packages such as contouring and three-dimensional operations are available.

- A debugger called NDB. This is similar to the standard UNIX tool called dbx, and provides source level debugging. As this uses the standard message-passing primitives, it can be used to debug parallel programs.

- A performance evaluation tool called PM. This is a graphical system providing the user with feedback on the way in which the transputers are being used. The user can identify various potential performance bottlenecks such as CPU usage, message-passing latency and link traffic.

Express also provides a library of functions which encompass a large part of the underlying work required to construct a parallel application. Consider a large job where the same action is to be performed on a number of pieces of data, such as image enhancement of a large picture. The normal style of parallel programming used here is to create a master program and many copies of a slave, each running on a separate transputer. The master program divides the work into slices which are handed out to the slaves, each of which performs the work in parallel.

Express provides a number of useful tools for this purpose. The first is a utility that generates runtime parameters for the slave programs, allowing them to adapt to the number of nodes and other changes in their environment. The second tool may be used to create the master program automatically, mapping the large picture into smaller pieces suitable for distribution. A third tool combines the output from the slaves back into a single answer.

The library routines can also be used to disguise the number and topology of the transputer network. For example, a routine is provided which is similar to the C routine `scanf`. When the user types a line of input at the terminal every slave is sent a copy of the line, just as if the line had been typed to each slave. In the same way each slave can issue an output command such as `printf`, but only one of the identical copies will be printed on the screen.

The library also provides explicit support for processor farms. A farm is a useful technique when the same subroutine must be run on a large number of data items and there is no intercommunication between the different calls. Express provides support for the model where the majority of the program code still runs on the host machine. A small change is made so that the program on the host specifies what data are to be sent to the transputer network, what results are expected back and what subroutine to call in parallel on the data array. The programmer need not be concerned with the actual communication required in order to package the data and transmit it around the network.

6.2 Linda

Linda is another system that has been designed to make it easier to program in parallel. It can be described as a programming language, although it is actually a set of calls embedded within an existing programming language such as C or FORTRAN, providing some of the functionality of an operating system.

Linda was designed by David Gelernter (1985,1988) in the Department of Computer Science at Yale University as a portable parallel programming system that was both efficient and easy to use. The transputer implementation is just one of a large number of different ports of Linda which range from shared memory systems to a network of minicomputers.

The basic abstraction of Linda is very simple and rather elegant. Linda supports the idea of data represented by 'tuples', which are a sequence of data fields each of a defined type. These tuples may be launched into 'tuple space' from which they may be either copied or extracted. Tuples are taken out of tuple space by matching field values and not by specifying some tuple identifier. This latter is familiar to users of databases, whereas tuples in Linda can only be retrieved by performing some suitable SELECT operation on tuple space.

This basic idea of Linda is easy to understand. Tuples are thrown into the tuple space rather like fish into a pond. Any processor can throw fish into the pond, and any processor can place a fishing rod into the pond. Each rod has a value attached to it, and only tuples that match the value will be caught. The template used as the basis for the match may specify any combination of the fields of the tuples, and tuple values may include a type of wild card which will match any template value.

This is actually implemented by providing a set of function calls in the host language, as shown for C in Table 6.1. This mechanism is good for the type of parallel problem described above which can be divided up into slices to be handled by multiple copies of the same slave program. The master simply starts up the slave processes by using the `eval(t)` call. This causes a new process to be spawned to evaluate the tuple passed to it. This tuple will contain the initial parameters for the program. The master then divides up the problem and passes the slices into tuple space using `out(t)`. Each slave process waits for work by using `rd(s)` to retrieve any tuple. Alternatively a slave may wait for some special tuple because it can handle certain types of work more efficiently.

Linda provides a useful abstraction for parallel programming. The most obvious potential flaw is the efficiency of the associative memory mechanism required to implement the tuple matching. On a static topology a compiler can be used to direct tuples to likely candidates, but even here some data tuples may have to be passed around redundantly between processors. In a more dynamic system the inter-processor communication bandwidth of transputers may easily be exceeded.

Table 6.1 Linda function calls in C

Function call	Description
`out(t)`	evaluate tuple **t** and then add it to tuple space
`eval(t)`	add tuple to tuple space and then evaluate it
`in(s)`	wait until tuple matching template **s** is available, then extract it
`rd(s)`	wait until tuple matching **s** is available, then extract a copy
`inp(s)`	if tuple matching **s** is available extract it, else return false
`rdp(s)`	if tuple matching **s** is available extract copy, else return false

6.3 Trollius

One of the approaches to the problem of running UNIX on a transputer system has been that taken by Trollius. This is an operating system for parallel architecture computers where each computer is regarded as consisting of a number of transputer-based calculation nodes connected to some conventional architecture machine running UNIX. Originally named Trillium, Trollius was designed and implemented within the Cornell Theory Center.

Trollius consists of a kernel that is booted into a network of transputers, which provides a set of library calls for programs running under it. Two languages, C and FORTRAN, are provided, as well as a number of development tools such as an assembler, linker and debugger. The system is designed to extend the existing UNIX environment of a single user into a set of transputers which then act as an accelerator. Support is provided for programmers in C and FORTRAN to access the facilities of the UNIX host such as terminal and file systems. A message-passing scheme is implemented which allows UNIX and any transputer node to exchange messages, and for any node to communicate with any other.

Trollius can be seen more as an extension to an existing UNIX system rather than a complete implementation of UNIX. There are very few of the UNIX system calls supplied within the transputer, no support for multiple users within a transputer network connected via a single link adaptor and no user interface or user level commands running on the transputer itself.

6.4 Mach

Mach is a multiprocessor operating system originally developed at Carnegie Mellon University, and currently being expanded and enhanced by a large number of commercial companies (Accetta *et al*, 1986). It was designed to provide a new foundation for UNIX development by providing access to both loosely coupled and tightly coupled processors in a coherent fashion. The original development was for the VAX and attempted to maintain binary compatibility with the standard Berkely UNIX 4.3 bsd.

To our knowledge there is no current implementation for the transputer, and indeed this is not surprising given Mach's support for the large virtual address space, which cannot be supported by the current range of transputers. However, a brief discussion of Mach is included here because the style of Mach is highly relevant to the transputer, and because future releases of the transputer may be able to support it.

One of the main features of Mach is the distinction between tasks and threads; a task is a unit of resource allocation and includes a unique address space. Each task contains multiple threads that share the address space and may execute in parallel on different processors.

Virtual memory is provided for each task, and any thread may access any part of the virtual address space. This essentially provides a shared virtual memory between threads within the same task. The sharing of virtual memory could lead to unacceptable performance on a system of connected transputers, but is clearly useful on a machine with true shared memory.

Mach provides a capability-based inter-process communication scheme based on messages containing typed data. These messages are sent across machine boundaries in a transparent fashion, requiring no specific network support. In order to maintain a consistent client–server interface an interface language is used. The message-passing system is integrated with the virtual memory mechanism so that large messages are sent efficiently using the physical shared memory when available.

6.5 Meikos

One of the earliest designers of hardware based around the transputer has been a company called Meiko. Their highly successful 'Computing Surface' was designed from the outset to use transputers as the engine for a scientific and engineering supercomputer (Chesney and Ganz, 1989). That the Computing Surface should use transputers is hardly surprising since Meiko was set up by some of the original designers of the transputer chip after they left INMOS.

The Meiko solution to software systems for the Computing Surface has been to provide a mechanism whereby the different users can share the Computing Surface resource. This is done by dividing the computer into a number of 'domains', each of which has no access to any of the others. The division into domains, and the associated inter-domain communication, is handled by a combination of system software and dedicated hardware.

Within each domain the user is free to run whatever software he or she wishes. The entire domain may be used to run a program intended to run on naked transputers. The program can communicate with the host using various server protocols, and communication from the domain to the actual server is handled by the underlying system software. All the links are configured electronically, so that it is possible to order up different domains in different topologies and test out varying parallel programming solutions quickly and easily.

The domain approach also allows operating systems to run within a domain. Meiko provide a version of UNIX called Meikos, which is essentially a single-user system. Multiple copies of Meikos can run for many different users, each in its own domain. A separate domain runs a fileserver providing shared file access.

This is a highly flexible approach, but it still requires the user to be aware of such issues as transputer topologies and parallel programming techniques. Although it is easy to order a larger domain than the one currently being used, the source of the program must be available so that it can be recompiled for the extra processors involved.

Meikos itself runs in a single processor; in order to use the extra processors within the domain an application written in a suitable parallel language must be used.

6.6 Helios

The operating system called Helios was designed by Perihelion Software for multiple-processor transputer systems (Perihelion, 1989). Although appearing similar to UNIX at the user level, the underlying implementation is entirely different in order to handle the multiple processors. Helios is based on the client–server model for operating systems, a technique that is widely used in many current systems, such as NFS and X Windows. A client process wishing to access a system resource, such as opening a file, sends a message to a server process requesting this action to be performed on its behalf. The server replies with another message indicating success or failure. Subsequently the client can read or write this file by sending further messages to the server. This inter-process message passing is used to provide a model for parallel programming that is independent of the underlying transputer architecture.

In a standalone transputer network a parallel program consists of communicating processors. Each processor runs a number of processes, and processors communicate by writing data down links.

Under Helios, programs consist of communicating tasks. Each task consists of a number of processes, and tasks communicate by reading and writing data down pipes. The actual mapping of tasks to processors is left up to the operating system, which also handles the multiplexing of physical links. Although a task may contain multiple processes, a task may not straddle processor boundaries. More than one task may run within a single processor.

The pipe mechanism is implemented on top of the underlying message-passing system, thus allowing the tasks making up the parallel program to be distributed over an arbitrary network of interconnected processors. The program must still be constructed using suitable parallel algorithms, but a binary version can be provided which runs on any number of processors connected in any topology. Of course, such a program will run faster on certain systems than others.

Consider the example of a master program and several slaves, where the master allocates work to the slaves as they become ready, and each slave sends back an acknowledgment when the work has been completed. This could be represented as the logical model given in Figure 6.1, where there are four slaves and a single master.

This logical description is translated by Helios into a physical mapping given the processors available. For example, if only two processors were available, then the system might look as shown in Figure 6.2.

Alternatively, if many processors were available, each task might be allocated a processor all to itself, as shown in Figure 6.3.

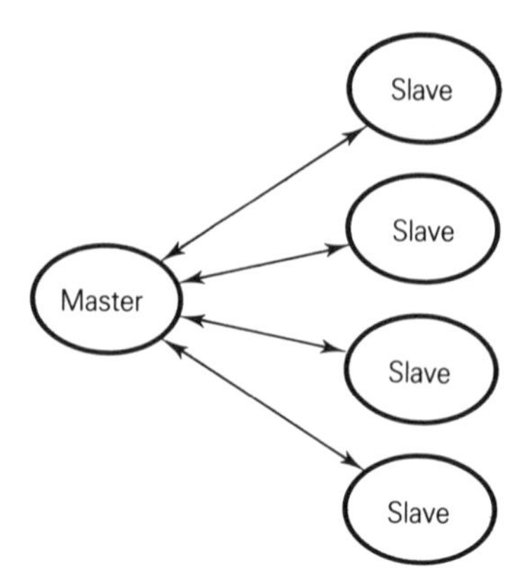

Figure 6.1 System of a master task and four slaves

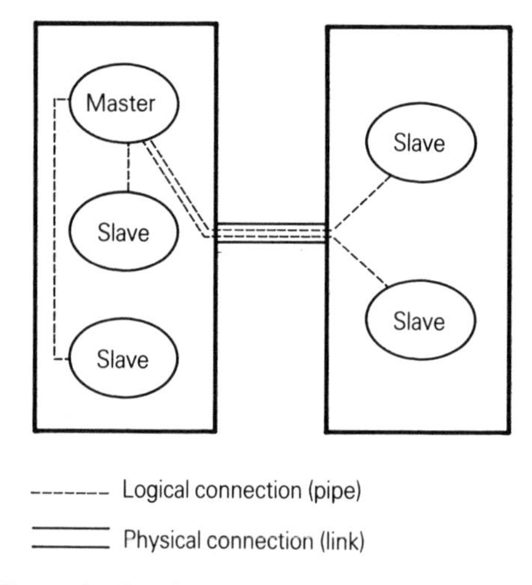

Figure 6.2 Five tasks distributed over two processors

The message-passing system and library I/O calls are handled by the Helios nucleus. This must be installed on each processor in the system and requires about 100 Kbytes of memory. The nucleus also provides memory allocation routines, semaphores and performance monitoring code.

It is useful to compare the Helios approach to parallel programming with the standalone languages described in the previous chapter. In a naked transputer network, communication is handled either by a language feature such as `?` or `!` in occam or by a library call such as `chan_out_word` in C. In Helios this is a standard library I/O call, such as the POSIX routines `read` and `write`.

The function of creating a parallel execution thread is handled by the `PAR` construct in occam. In standalone C this can be performed by the routine call

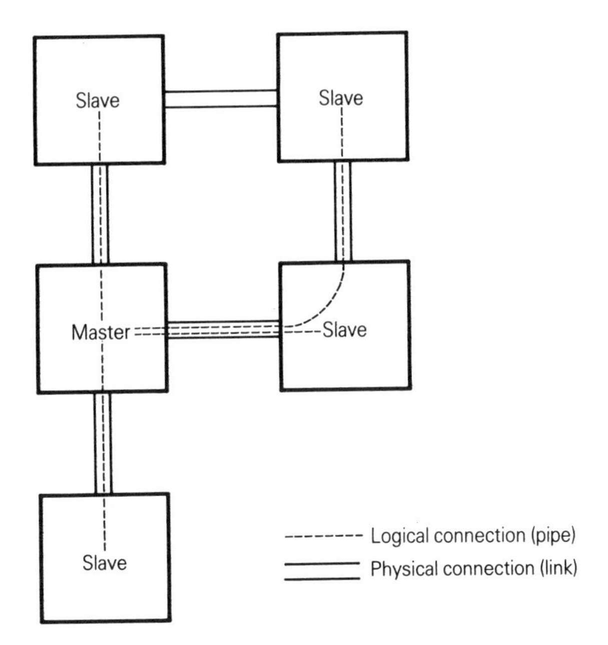

Figure 6.3 Five tasks distributed over five processors

`create_thread`. This is similar under Helios, where the `Fork` routine is used to create a thread (or process) within a Helios task running in the same processor.

A parallel thread that is to run on another processor must be identified to the configurer program before an occam or standalone C program is run. Under Helios the POSIX routine `vfork` is used to dynamically create a new task, possibly within a different processor.

The action of waiting for one of a number of events to occur, and then scheduling a process to run in order to handle that event is handled in occam by the `ALT` construct. Under Helios and in standalone C the events are allocated a thread each, and each thread waits for its event to occur. When an event takes place the corresponding thread sets a global value to indicate what is happening, and uses a semaphore to wake up a central thread. This is similar to the UNIX approach of using a select mechanism to await an I/O event on a number of different channels, although by having multiple processes the event can come from a timer, I/O or the transputer's `EventReq` input.

6.7 UNIX

In many transputer applications the UNIX operating system would seem to be the immediate and obvious choice. UNIX is widely used in workstations and in other scientific and engineering applications. There are a range of languages, other tools and complete applications ready to run under UNIX. It is widely supported by a large range of hardware manufacturers from the giants like IBM and DEC right down to tiny system integrators. It has been available for over ten years and the basic system is mature and resilient.

There is, however, one major problem with UNIX. This is that UNIX was originally designed for a single processor and a memory management unit. A good example of the classic UNIX computer was the PDP-11 from DEC, which might contain a 72 Mbyte disk and 512 Kbytes of RAM. Within this, UNIX would provide support for five or so user processes plus ten or so system processes, where each process was limited to 128 Kbytes. A paged memory management unit was used for two major operations. The first of these was to enable UNIX to roll out to disk pages of memory belonging to inactive processes. When a process was later scheduled and attempted to access this non-existent page, an interrupt was caused that enabled UNIX to swap in the page.

Secondly, the memory management unit provided every process with a private memory address space, each starting at zero and extending as large as required subject to the addressing constraints of the underlying processor. In this way the whole 512 Kbytes of memory in a PDP-11 could be shared by many processes, each no larger than 128 Kbytes. Perhaps more importantly the private address space provided protection between different processes. It was not possible for one program running as a UNIX process to read or write the memory associated with any other.

This inter-process protection ensured that UNIX was popular in many academic institutions, and from there its popularity has spread. Modern implementations normally use advanced memory management units to provide true virtual memory, giving every UNIX process an enormous virtual address space of 16 Mbytes or so, expanding as required. Memory protection is enhanced so that code areas may be shared but the individual data areas of each process are protected.

The transputer does not have memory management. There is none provided on the chip and it is practically impossible to add it externally, as individual instructions are not restartable. This was not an oversight on the part of the designers; the transputer is intended to be used in conjunction with other transputers where each processor has its own private memory. The original transputer design envisaged an operating system such as UNIX running on a network of transputers, where each process runs on its own processor. In this kind of system memory protection between processes is the same as that between processors; such a mechanism is completely foolproof and would not be subject to the same 'trapdoors' by which it is possible for malicious programs to corrupt other processes in shared memory machines.

There are problems with the one process/one processor approach to implementing UNIX. The first is that it is far too expensive, especially when many UNIX processes exist merely so that they can wake up every hour or so and create another process. In these circumstances it is highly wasteful to allocate a complete transputer and external memory just for this job.

The second problem is that there is still no virtual memory. If a program needs an address space of 16 Mbytes, it needs 16 Mbytes of physical memory connected to it. If UNIX wishes to run ten user processes, each with a process address space of 16 Mbytes, then at first glance ten transputers each with 16 Mbytes will be required.

In theory it is possible to share the memory in a multiple transputer system by extracting the code and data of a process from a particular transputer, saving this on disk somewhere and then starting a new process in the transputer with the large memory attached. When a suitable time slice has expired, the running program could be halted, extracted and written to disk and the original program reinstated. This approach would allow several large processes to run on a transputer system where there were only a few large memory processors, although the cost of swapping between processes would be high. We know of no current implementation that attempts to do this. The maximum size of any process would still be limited by the maximum amount of external memory connected to a processor in the network.

A third problem concerned with implementing UNIX is that the transputer already contains many of the mechanisms normally found inside operating systems. In particular, UNIX contains a kernel responsible for creating and manipulating processes and for communication between processes. The transputer already has much of this in its hardware and microcode. A UNIX implementor may decide to ignore this support and implement another type of process mangement system, thus wasting the support provided and creating a much slower and less efficient alternative.

On the other hand, the implementor could decide to delve into the internals of UNIX, attempting to match the transputer's view of a process with that of UNIX. Besides the problems involved with retaining compatibility and reliability, UNIX expects the processor to be able to be switched to a 'supervisor' mode where it can inspect the state of processes and internal data structures without user processes running. This is only possible using transputer processes if the 'supervisor' state is implemented by switching to a high-priority process. This priority is also used for interrupts so that the interrupt handling latency is extended and the system no longer reacts in real time.

An alternative approach to providing a UNIX-like operating system for the transputer has been taken in a system known as transIDRIS. The original IDRIS was a UNIX-like system, similar to UNIX Version 6, which required no memory management. IDRIS has since been extended to provide a set of POSIX-compatible calls within an operating system that has been ported to a large number of different processors. The transIDRIS system is the name given to the port to the transputer.

In transIDRIS the central transputer is taken to be the master, and all extra transputers are regarded as slaves. Within the master processor runs a nucleus that contains all of the functions normally provided in a POSIX system. Each slave processor contains a cut-down kernel which provides inter-processor communication back to the master processor. A process running on a slave processor makes system calls that are actually handled by the master processor, with the slave kernel handling the communication back to the master.

In many ways there are similarities here to the Trollius system, except that in the case of Trollius the calls are passed back to some other processor running UNIX, and in transIDRIS the calls are passed back to a master transputer running transIDRIS.

Another alternative approach has been taken by a US company, Cogent Research. They have produced a transputer workstation that uses Linda, with PostScript to handle the graphics. This is all implemented on top of a simple UNIX-like system that is similar in many ways to the Minix operating system described by Andrew Tanenbaum (1987). Here the inter-process message passing has been arranged so that messages can be passed transparently between processors. A function known as the 'oracle' is in charge of the way in which processes are allocated to processors.

Recent announcements from INMOS would seem to indicate that future versions of the transputer will provide some form of memory management. It is unlikely that this will include a full-paged virtual memory system, at least in the short term. However, in the long term changes to the underlying hardware may well enable the porting of standard operating systems, such as UNIX System V.

Chapter 7

Transputer Family Hardware

In this chapter we begin the discussion of the hardware aspects of the transputer. Although each device has its own particular characteristics there are sufficient similarities, even at the input and output signal level, for it to be useful to discuss transputers as a family, and to consider the differences between devices only when necessary. As a typical example, Figure 7.1 shows a block diagram of the T805, one of the most recent transputers, and Figure 7.2 the data paths within the T805. This will be used as a model against which to compare the other transputers.

7.1 Hardware architecture

The register model of the transputer and its instruction set have been discussed in detail in Chapters 2 and 3. All transputers have a fast integer processor, with an instruction set that has evolved over the past few years. Many instructions operate in a single cycle of the processor clock, and transputers are at present manufactured with clock speeds of up to 25 MHz, with 30 MHz devices expected shortly. INMOS have adopted a fairly conservative approach to performance figures, claiming 10 Mips (million instructions per second) for a 20 MHz processor. All transputers operate from the same external clock speed of 5 MHz; the processor clock is obtained from an internal phase-locked loop multiplier.

The processor fetches instructions a word at a time, in order to reduce the effect of the lower access speed when code is kept in external memory. However, there is no data cache, and data fetches will be slowed if data are in external memory.

7.1.1 Floating-point unit

The floating-point unit of the T8 series transputers is a 32/64-bit unit, conforming to the IEEE 754–1985 specification. The FPU contains an evaluation stack similar to that of the integer processor, with three registers *FA, FB* and *FC*. Each register can contain either a 32- or 64-bit number, and has a flag to show which is stored. Its design is the result of a compromise between maximizing overall processor performance and minimizing chip area. Thus the FPU has no flash multiplier or barrel shifter. Despite this, the performance is quite satisfactory, with single and double precision multiplication times of 550 and 1000 nsec respectively, for a 20 MHz device. The FPU operates concurrently with the integer processor, and thus computation can be speeded up by overlapping integer and floating-point processing. This is especially important in compiling a language such as FORTRAN, where the cost of array index calculations can almost be hidden by clever code generation.

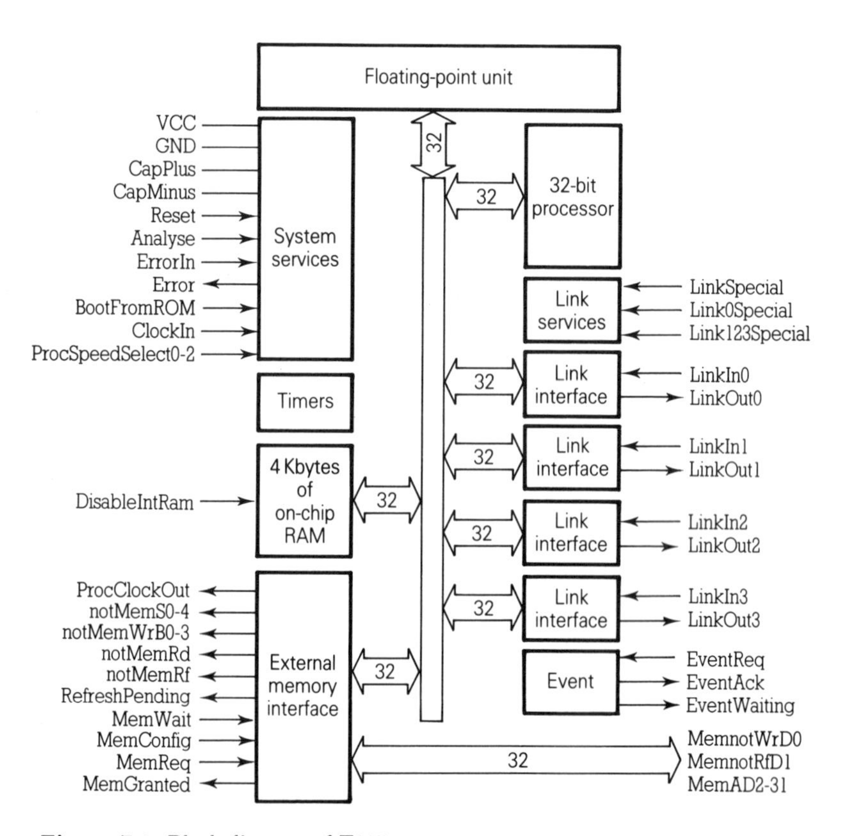

Figure 7.1 Block diagram of T805 transputer

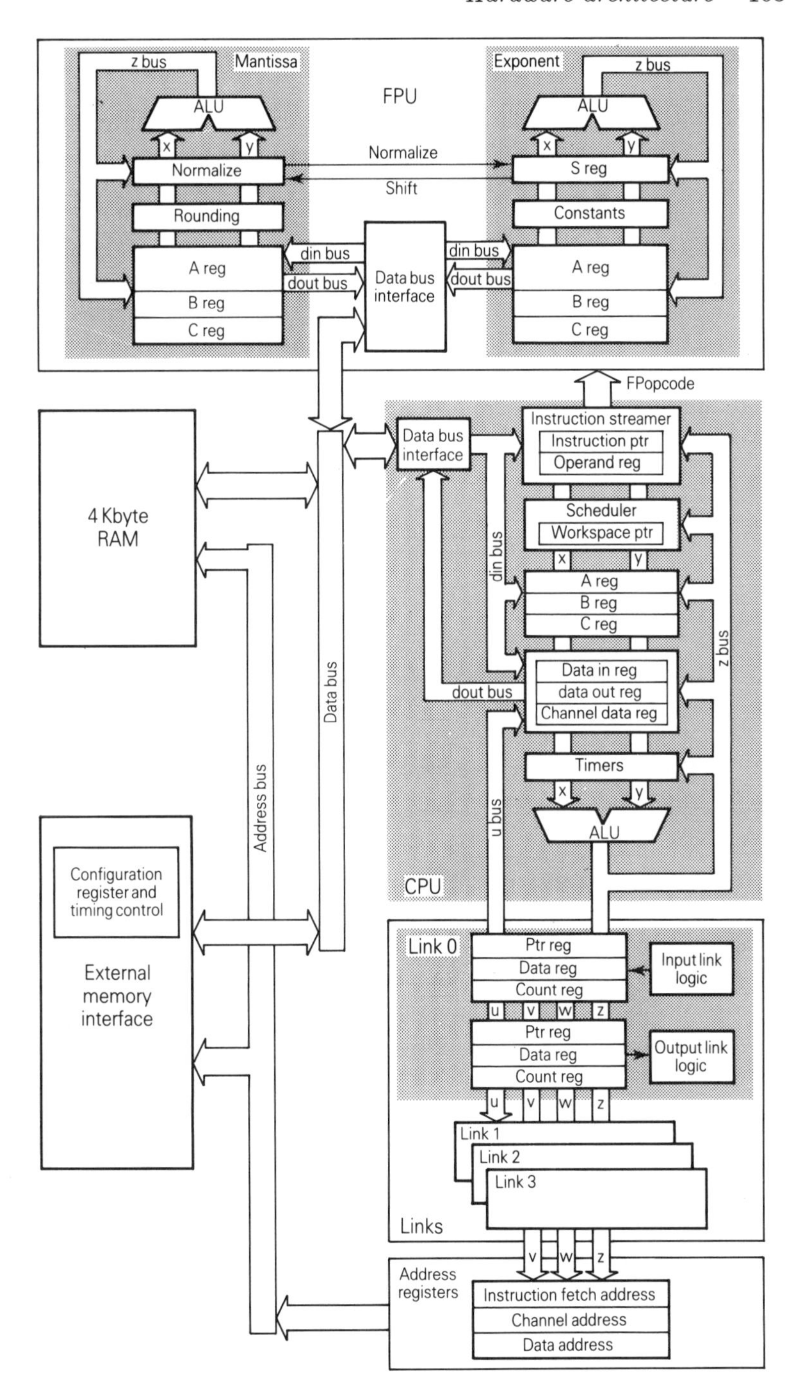

Figure 7.2 Data paths in the T805 processor

7.1.2 Links

The INMOS link is perhaps the key to the entire transputer system. A link will provide a simple two-wire connection between transputers that is capable of transferring data at almost 2 Mbytes per second. The link operates at TTL levels, and direct connections can be made for distances up to a few inches, suitable for connection within a circuit board, or between adjacent boards. For longer distances, the links can be made by matched transmission lines, driven by more powerful buffers, or transmitted along optical fibers. The link protocol provides no error detection or correction capabilities, thus requiring that the physical medium that carries the data be error free, or that errors be handled by software. Three standard speeds of 5, 10 and 20 Mbps are supported, although not all transputer family devices can work at all of these speeds.

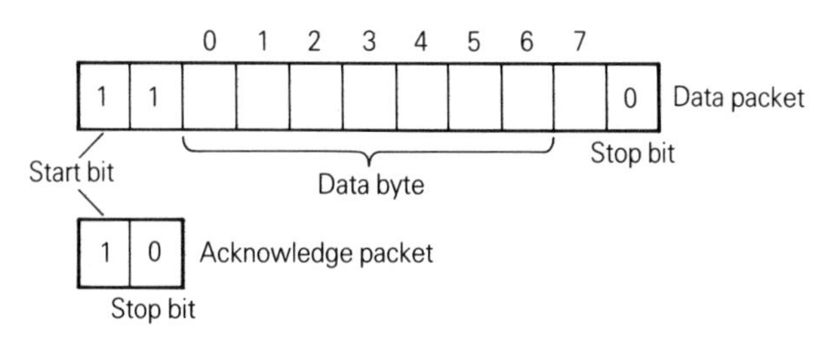

Figure 7.3 Link byte format

Data on the link are sent one byte at a time, and each byte must be acknowledged by the receiving transputer. This acknowledgment indicates to the transmitting device that it can send another byte. The byte format is shown in Figure 7.3; a data packet consists of a high start bit, followed by a second high bit, and then the eight bits of the data byte. It is terminated by a single low stop bit. The acknowledge packet has a high start bit, followed by a single low stop bit. In early transputers, such as the T212 and T414, the acknowledgment is not sent until the entire data packet has been received. Thus a minimum of 13 bit times, or 1.3 μsec at 10 Mbps, is required for each byte to be sent and acknowledged. This gives a theoretical upper limit of 751 Kbytes per second at 10 Mbps and 1.47 Mbytes per second at 20 Mbps.

The T800 sends the acknowledge packet soon after it has started to receive the data packet. Thus only 11 bit times are required per byte, giving data transfer upper limits of 888 Kbytes per second at 10 Mbps and 1.73 Mbytes per second at 20 Mbps. This is a substantial improvement over the T414, and overlapped acknowledge has been implemented in all the more recent transputers. Table 7.1 shows measured data transfer speeds between various processors at different link speeds. This has been measured with a message size of 64 Kbytes, so that the overhead due to the transfer setup time is negligible.

Table 7.1 Measured unidirectional link speeds

Transfer	T800→T800	T414→T800	T800→T414	T414→T414
Link speed	1.69	1.36	0.90	0.80
Bit times per byte	11.3	14.0	21.0	23.8

Measured with T800c and T414b transputers at a link rate of 20 Mbps, in units of Mbytes per second

It can be seen that the actual throughput achieved is lower than the theoretical maximum, especially for transfers that do not use overlapped acknowledge. Data rates can also be reduced by delays in the physical link, which lengthens the time taken for the handshake; this is discussed further in Section 7.4 and Chapter 8.

Each of the presently available transputers has four external links, except for the M212 where two of the links are used in the disk interface, and the low-cost T400 where only two links are provided in order to save silicon real estate. The link data are transferred between the link and memory by DMA, with no processor intervention after the transfer is set up, and therefore communication uses almost no processor time. The traditional method used to connect transputers over short distances is with an unscreened, twin-twisted pair cable. This results in a link with a characteristic impedance of about 100 ohms. The quality of the TTL signal on such a link gradually deteriorates as longer links are used. Eventually, after 20 inches or so, the signal becomes too distorted, and errors will occur. As mentioned above, the INMOS link assumes that the transmission medium is error free. Methods of connecting links over longer distances are discussed in Chapter 8.

7.1.3 Memory

All transputers have a small amount, 2 or 4 Kbytes, of internal static RAM. This is very fast, and accessed in one cycle of the processor clock. Thus the bandwidth of the internal memory varies from 40 Mbytes per second for a 16-bit device running at 20 MHz, to 120 MBytes per second for a 30 MHz, 32-bit device. This memory occupies the lowest portion of the address space, where some locations are reserved for processor functions. On some processors the internal RAM can be disabled by applying a high level to the `DisableIntRam` input. When this is done the corresponding portion of the address space is mapped to external memory. Of course the processor will not function if this external memory does not exist. As the internal memory is seldom sufficient for applications programs, each transputer has an external memory interface or EMI. The 16-bit processors have a non-multiplexed external address and data bus which can address up to 64 Kbytes of memory. This interface is optimized for static RAM, and can read and write in two processor cycles. Thus external memory is just half the speed of the internal RAM, giving a maximum memory bandwidth of 20 Mbytes per second for a 20 MHz device.

The T801 has a very similar external memory interface, with 32-bit address and data buses. This is capable of data transfer rates of up to 60 Mbytes per second for a 30 MHz device.

The T4 series processors and the other members of the T8 family have a more complex multiplexed EMI. This is programmable and can generate all the signals necessary to control dynamic memory devices. Several preprogrammed configurations can be selected, or the interface can be programmed by a string of 36 bits read from high memory after the processor is reset. The complexity of the interface makes it slightly slower, and an external memory access requires at least three processor clock cycles. However, it is more common for large memory designs to require four or even five cycles unless very high speed memory devices are used. The topic of designing systems with the transputer memory interfaces is discussed in detail in Chapter 9.

7.1.4 Error signals

Internally transputers have an *ErrorFlag* flag that is set by certain error conditions. The `Error` output reflects the state of this flag, `OR`'ed with the state of the `ErrorIn` signal, which is provided only on the T805, T800 and T425 processors. If the *HaltOnError* flag is set, the processor will halt as soon as *ErrorFlag* is set. On early transputers the *ErrorFlag* can only be reset by executing the `testerr` instruction as *ErrorFlag* is not cleared by processor reset, thus making it possible to detect this condition after reset even when `Analyse` has not been asserted. However, in the T805 and T425 *ErrorFlag* and *HaltOnError* are cleared on reset if `Analyse` has not been asserted.

7.1.5 Event handling

Transputers have only a single source of external interrupts, the `EventReq` input. This signal is active high, and when asserted makes the external event channel ready to communicate with a process. If there is a process waiting for input on the `Event` channel, the processor takes `EventAck` high, and the process is scheduled. `EventAck` is taken low by the processor a maximum of one processor cycle after `EventReq` is returned low. If the waiting process is at high priority it will run when any other high-priority processes scheduled before it have completed or are waiting. Similarly, a low-priority process will run when all high-priority processes have completed, or are paused, and any other low-priority processes scheduled ahead of it have either had their time slice or are waiting for I/O. Thus, for predictable response to interrupts the event response process should be run at high priority and be the only high-priority process running. If this guideline is followed, then the interrupt latency – the time before the high-priority process starts running – is typically 19 processor cycles, and a maximum of 78 cycles, assuming that on-chip RAM is being used. If the FPU is not being used, then the maximum time is

58 processor cycles. For a 20 MHz processor the interrupt latency is thus typically 950 nsec, with a maximum of 2.9 μsec if the FPU is not being used and 3.9 μsec if it is.

The T805, T801 and T425 processors also have an `EventWaiting` output. This is asserted whenever a process attempts input from the `Event` channel. Thus this output can be used to indicate to a peripheral that the transputer is ready to respond to an `EventReq` input. `EventWaiting` is taken low as soon as `EventAck` goes high.

7.1.6 Reset, analyse and bootstrapping

As with most microprocessors, transputers have an external reset signal that returns the device to a known state after switch on. However, an `Analyse` input is provided that, if asserted before `Reset`, causes the transputer to save some of its internal state. This can be a very helpful debugging tool.

The timing of the `Reset` signal is shown in Figure 7.4. `Reset` must be high for a minimum of eight periods of `Clockin`, and the clock must be running for at least 10 msec before `Reset` goes low again.

The `Analyse` input acts as a debugging aid. If `Analyse` is taken high the processor will halt and save some of its internal state. If the processor is then reset, the saved state can be examined using peek and poke operations down the links. The time taken for the processor to halt depends on the processes running when `Analyse` is asserted, as the processor halts at the next descheduling point of a low-priority process. This can be as long as three time slice periods after `Analyse` goes high.

The actions taken when `Reset` goes low depend on the type of memory interface and the state of the `BootFromRom` input. The programmable interfaces of the 32-bit transputers require a configuration period, as described in the memory interface section above; this is not required by the 16-bit transputers or the T801. After memory configuration the transputer will bootstrap itself from memory if the

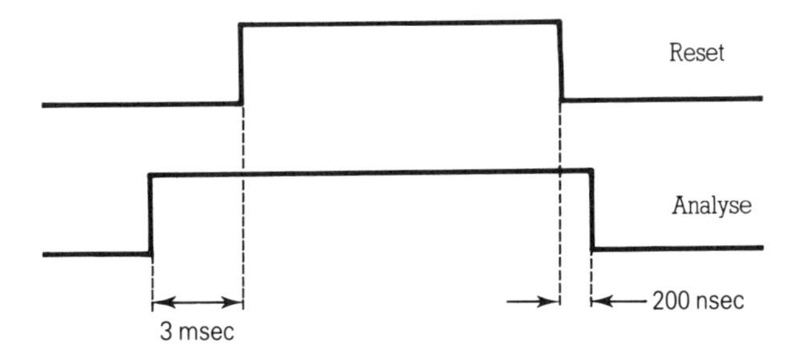

Figure 7.4 Timing of `Reset` and `Analyse` signals

`BootFromRom` input is high. The bootstrap is assumed to lie at the most positive end of memory, at `0x7FFFFFFE` for 32-bit transputers and `0x7FFE` for 16-bit parts. This location would normally contain a jump to bootstrap code in some other part of the memory map. The possible contents of bootstrap code are discussed in Chapter 4.

If `BootFromRom` is low the transputer will wait for messages on its links. If the first or control byte of the message is greater than 1 then that number of bytes are read from the link into memory, starting at location `MemStart`. This code is then started as a low-priority process. It would normally control the reading in of a secondary bootstrap, which would load the applications program or operating system. If the control byte is 0, the transputer expects two more words of data (16 or 32 bits, depending on the processor). The first word is interpreted as a memory address, and the contents of the second word are written (poked) into memory at that address. If the control byte is a 1 then the transputer expects a single word from the link, which is also interpreted as a memory address. Data are read (peeked) from memory at this location and transmitted down the link. Any link can be used for the peek and poke process, but each complete operation must use the same link. At any time the receipt of a message with a control byte of greater than 1 will cause the transputer to boot from that link.

7.1.7 Clocking

The `ClockIn` input expects a square-wave clock signal at 5 MHz. The clock frequency is independent of the transputer type and processor speed; internal clocks are generated by a phase-locked loop frequency multiplier. The accuracy required of the clock is mostly a function of link communication between transputers; this will not function properly if the difference in clock speeds is more than about 400 ppm.

The processor clock signal is output on the pin `ProcClockOut`. On the more recent transputers the multiplication factor, and thus the processor frequency, can be selected by voltages applied to the three `ProcSpeedSelect0-2` pins, as shown in Table 7.2. Of course, the processor must be a suitable type to run at the processor

Table 7.2 Processor speed selection

Proc Speed Select2	Proc Speed Select1	Proc Speed Select0	Processor clock speed (MHz)	Processor cycle time (nsec)
0	0	0	20.0	50.0
0	0	1	22.5	44.4
0	1	0	25.0	40.0
0	1	1	30.0	33.3
1	0	0	35.0	28.6
1	0	1	invalid	
1	1	0	17.5	57.1
1	1	1	invalid	

speed selected. The `ProcSpeedSelectn0-2` pins are marked 'Hold to Ground' in the specification of the T414, which normally runs at 20 MHz. This is the only significant difference between the pinouts of the T800 and the T414, and means that a 20 MHz T800 is a pin-compatible replacement for a 20 MHz T414.

7.2 The transputer processors

The development of the transputer processors has been a process of gradual refinement of the instruction set and the external interface, and the correction of design errors. This development has been paralleled by a gradual speeding up of the processors, and a reduction in cost. However, these changes have been made in a way that has preserved upwards compatibility of both hardware and software.

The refinement of the instruction set has seen the addition of quite complex instructions for block moves and CRC calculations, as well as simple operations like `dup`, which duplicates the value at the top of the stack. The most recent transputers have a set of instructions for supporting debugging; these instructions will prove extremely useful as debugging transputers has been a very difficult process.

The hardware changes have both speeded up the processors and added useful signals such as `RefreshPending`, `EventWaiting` and processor speed selection. These changes have been made in a way that preserves pin compatibility with the earlier T414 and T212 transputers.

7.2.1 The 16-bit transputers

The T212 was the original 16-bit transputer, with 2 Kbytes of internal static memory; it has now been replaced by the T222 (Figure 7.5) which has 4 Kbytes of internal memory and an extended instruction set. The T225 is a recently announced version of the T222 with additional debugging instructions. The M212 disk processor (Figure 7.6) is a T212 with a built-in disk interface, 1 Kbyte of RAM and 4 Kbytes of internal ROM containing disk controller firmware. The M212 has only two external links as the other two are used as part of the disk interface.

The three 16-bit transputers have similar external memory interfaces, described in detail in Chapter 9. The internal memory occupies the most negative region of the 16-bit signed address space, starting at `0x8000`. The lowest locations are reserved for link and event channels (Table 7.3), but the space above these is available to users.

External memory space starts at the top of the internal memory and extends to the most positive address, `0x7FFF`, allowing 62 Kbytes of external memory on the T212 and 60 Kbytes on the T222. The amount of external memory that may be used with the M212 depends on the mode in which it is run. In Mode 1 the 4 Kbyte internal ROM is mapped into the address space at `0x7000` to `0x7FFF`, and

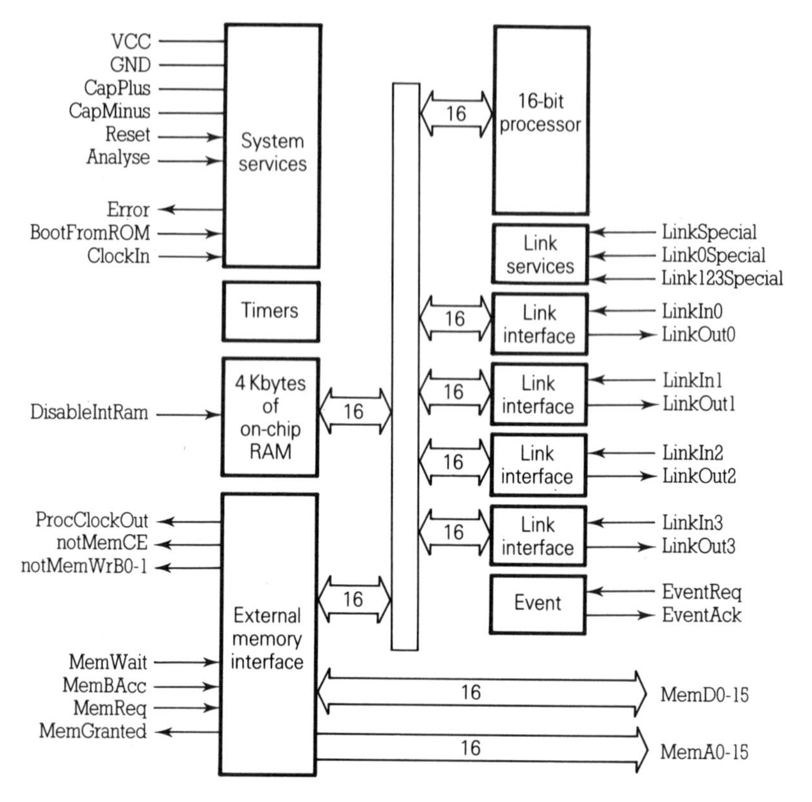

Figure 7.5 Block diagram of T222

only 58 Kbytes of external memory may be used. In Mode 2 the internal ROM is not available, and a full 62 Kbytes of external address space is available (see Table 7.3).

Both internal and external memory are 16 bits wide. The internal memory is accessed in one cycle of the processor clock, or 50 nsec for a 20 MHz device. This gives the internal memory a bandwidth of 40 Mbytes per second. The external memory interface requires a minimum of two processor cycles for each word transferred, thus an external memory access requires twice as long as an internal access.

7.2.2 The 32-bit transputers

The T414 (Figure 7.7) is the original member of the transputer family. It has a 32-bit integer processor, with 2 Kbytes of internal RAM. This is now being replaced by the T425, which has 4 Kbytes of internal RAM, the extended instruction set and debugging instructions.

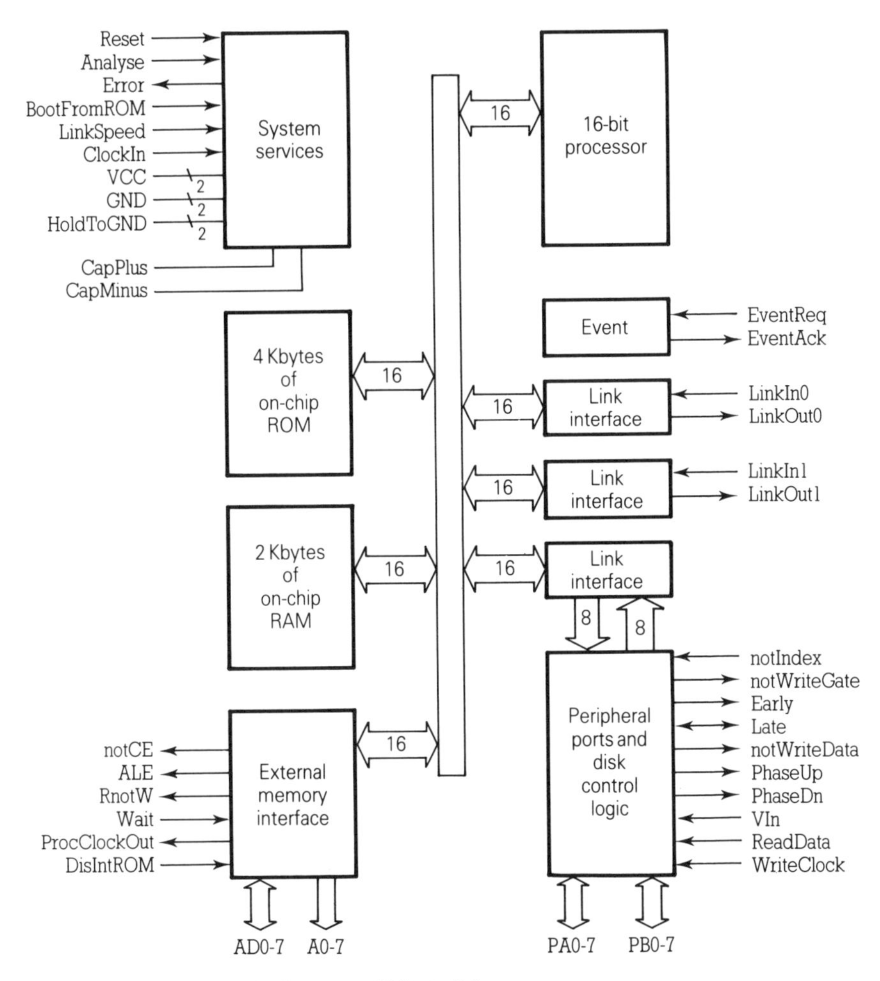

Figure 7.6 Block diagram of M212 disk processor

The first floating-point transputer was the T800, which also introduced both overlapped acknowledge and the extended instruction set. The T800 will eventually be replaced by the T805 (Figure 7.1) which has debugging instructions and adds the `EventWaiting` and `RefreshPending` outputs. The T801 is similar to the T805, but with a different memory interface. Each of these processors is available in processor speeds up to 25 MHz with 30 MHz versions expected soon.

The 32-bit transputers have a similar memory map to the 16-bit devices. Memory addresses are signed 32-bit integers, running from the most negative `0x80000000` to the most positive `0x7FFFFFFF`. Internal memory occupies the lowest part of the address space, and again the bottom locations are reserved for channel I/O, event and timer handling and register save areas. Extra space is reserved for the use of

Table 7.3 Memory maps for the 16-bit processors

	T212	T222	M212 Mode 1	M212 Mode 2
Base of internal memory	0x8000	0x8000	0x8000	0x8000
Start of user memory (MemStart)	0x8024	0x8024	0x8024	0x8024
Top of internal memory	0x87FF	0x8FFF	0x83FF	0x83FF
Start of external memory	0x8800	0x9000	0x8400	0x8400
Top of external memory space	0x7FFF	0x7FFF	0x6FFF	0x7FFF
Internal ROM				
Base			0x7000	
Top			0x7FFF	

the extended instruction set, and thus the start of usable memory (MemStart) is at 0x80000048 for the T414 and at 0x80000070 for the other 32-bit processors. Internal memory can be accessed in one cycle of the processor clock, giving a memory bandwidth of 80 Mbytes per second for a 20 MHz device, and 120 Mbytes per second for a 30 MHz device. The T801 has a non-multiplexed external memory interface, similar to that of the 16-bit processors. This requires a miminum of two processor clock cycles for each word transferred, half the bandwidth of the internal memory. The other 32-bit transputers have a programmable EMI, which contains a 10-bit refresh counter and can generate all the signals required to control dynamic RAM devices. This interface is rather slower than that of the T801, and requires at least three processor cycles for each access. However, three-cycle memory needs very fast and expensive memory chips and fast external address multiplexing, and most external memory implementations use four or five processor cycles to avoid these problems.

7.3 Transputer support devices

One of the important features of the transputer is that it does not need a wide range of support devices to provide such functions as communications and memory interfacing. However, INMOS have provided two styles of support device: the link adaptors to allow the connection of transputer links to more conventional bus-orientated systems and the link switch, which makes possible the electronic configuration of the links between transputers.

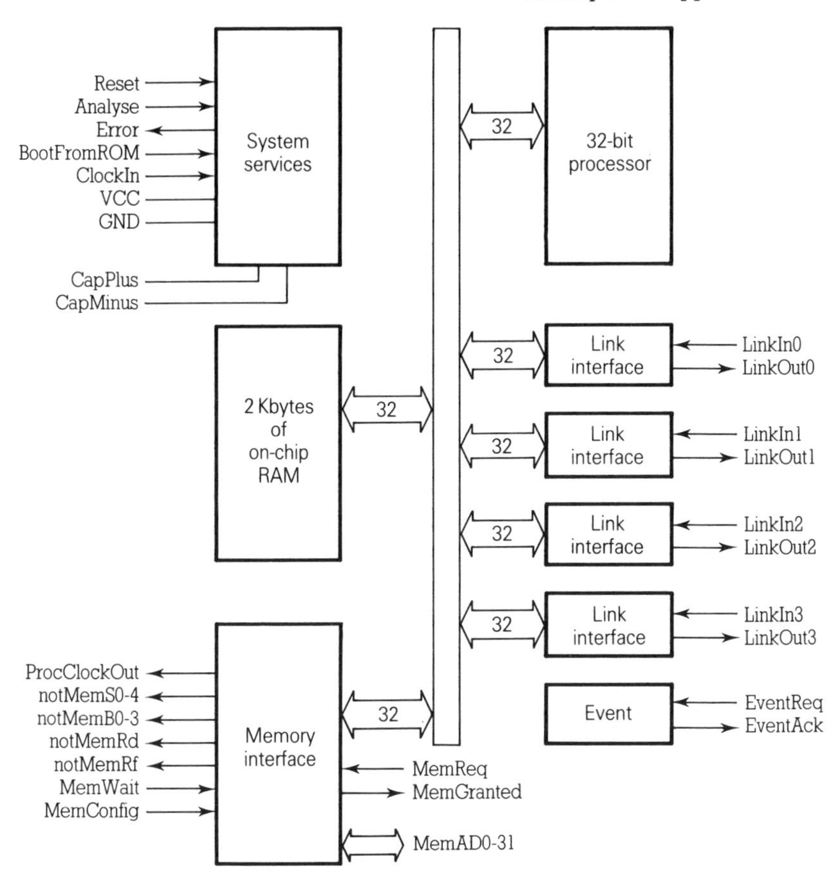

Figure 7.7 Block diagram of T414

7.3.1 Link adaptors

When it is necessary to connect a transputer link to a parallel device, such as the bus of a conventional microprocessor, a link adaptor may be used. Two link adaptor types are now available, the INMOS C011 and C012. Block diagrams of these devices are shown in Figure 7.8. The C011 has two modes: Mode 1 which provides separate 8-bit parallel data I/O ports; and Mode 2 with a single parallel I/O port. Mode 2 provides identical functionality with the C012 device.

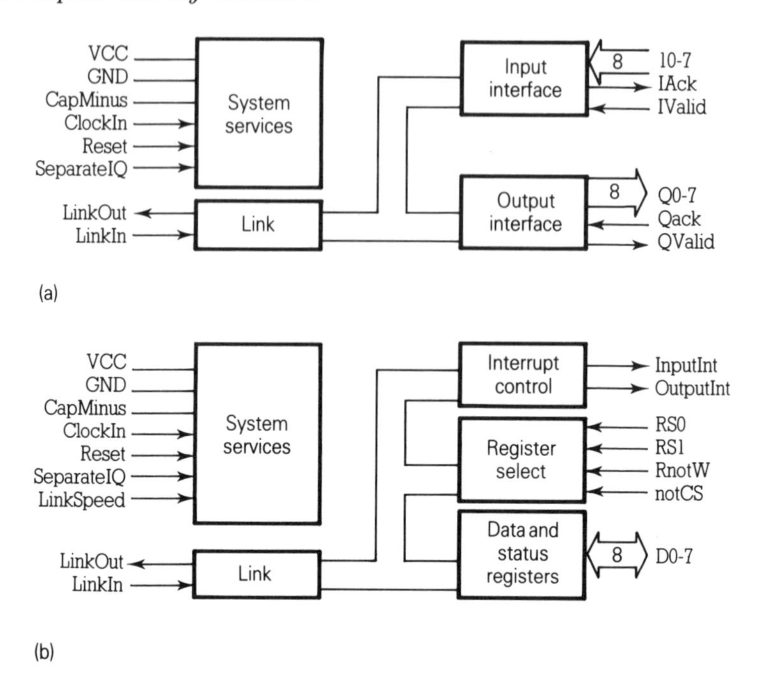

Figure 7.8 (a) C011 Mode 1 and (b) C011 Mode 2 and C012; input `SeparateIQ` does not exist on the C012

7.3.2 C011 Mode 1

In this mode the device is configured as a peripheral interface with separate data input and output ports (Figure 7.8a). The lines `IValid` and `IAck` provide a simple handshake on input to the device. When the data on `I0-7` are valid, `IValid` should be taken high by the peripheral device. The link then transmits the data on `I0-7`, and when the transfer is complete and has been acknowledged, takes `IAck` high. When the peripheral device takes `IValid` low, `IAck` is taken low by the C011.

A data byte received by the C011 is loaded into the output port `Q0-7`, and `QValid` is then taken high by the device. After the data have been read by the peripheral it must take `QAck` high. The C011 will then send an acknowledgment packet on the link to signal that data transfer is complete. It will then set `QValid` low, and the peripheral may take `QAck` low again. Table 7.4 summarizes Mode 1 C011.

Table 7.4 Mode 1 C011

Pin	I/O	Function
`LinkIn`	in	data from link
`LinkOut`	out	data to link
`I0-7`	in	data input from bus
`IValid`	in	indicates valid data written from bus into `I0-7`
`IAck`	out	indicates input data successfully transmitted to link
`Q0-7`	out	data output to bus
`QValid`	out	indicates valid data available in output register
`QAck`	in	indicates that output data has been read

7.3.3 C011 Mode 2 and C012

In Mode 2 the C011 has a bidirectional I/O port, and appears identical in function to the C012 (Table 7.5 and Figure 7.8b).

The device has four registers that are selected by the inputs `RS0` and `RS1`. The input data register contains the serial data input from the link. If there are valid data in the register, the data present flag will be set in the read status register (Figure 7.9a). This input data register is read only; writing to it will have no effect. The input status register has a data present flag (bit 0) and the input interrupt enable bit (bit 1) for `InputInt`. The data present flag is set when valid data arrives in the input data register. It is reset when the data input buffer is read, or by the `Reset` input.

Data written to the output data register will be transmitted down the serial

Table 7.5 C012 and Mode 2 C011

Pin	I/O	Function		
`LinkIn`	in	data from link		
`LinkOut`	out	data to link		
`D0-7`	in/out	bi-directional data bus		
`notCs`	in	selects the device		
`RnotW`	in	select reading from or writing to the parallel port, used with RS0-1		
`RS0-1`	in	selects one of the four registers:		
		RS1	RS0	Register
		0	0	read data
		0	1	write data
		1	0	input status
		1	1	output status
`InputInt`	out	interrupt on receive data ready		
`OutputInt`	out	interrupt on transmit data buffer empty		
`LinkSpeed`	in	select link speed: high – 20 Mbps, low 10 Mbps		

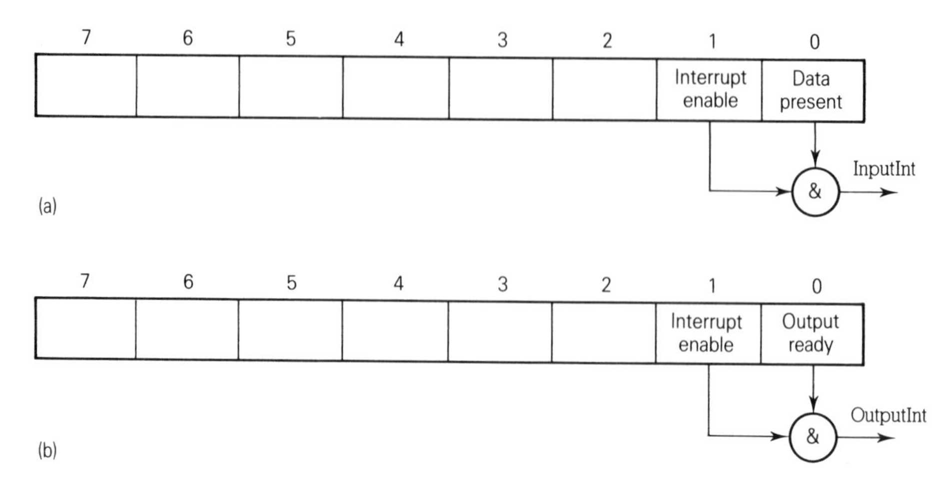

Figure 7.9 (a) C011 Mode 1 and (b) C012 status registers

link. The output ready flag in the output status register indicates that the device is ready for new data. Data will be lost if the device is written to when this flag is low. The output data register is write only. The output status register (Figure 7.9b) contains the output ready flag and the interrupt control bit for `OutputInt`. The output ready flag is set when the device is reset, and cleared when a data byte is written into the output data register. The flag is set again when the data has been successfully transmitted down the link and acknowledged. If the interrupt control bit is set, then `OutputInt` will be taken high.

7.3.4 The C004 link switch

Electronically reconfigurable transputer arrays are made possible by the use of the C004 link crossbar switch (Figure 7.10). This provides thirty-two inputs and thirty-two outputs, which can be connected in any configuration. The switch is programmed by a separate programming link. Like other devices of the transputer family, the links can run at either 10 or 20 Mbps, but 5 Mbps is not available.

As the bidirectional transputer link requires two wires, two inputs and two outputs from the C004 are required for each physical link. Thus a single C004 can set up sixteen connections between transputers. This allows all the possible interconnections of links in eight transputers, two link connections between each of sixteen transputers, and one link between each of thirty-two transputers. The internal structure of the C004 (Figure 7.10b) shows that each output has associated with it a 6-bit latch. The bottom five bits of this latch contain the number of the input to which this output is connected; the most significant bit indicates whether or not this connection has been made.

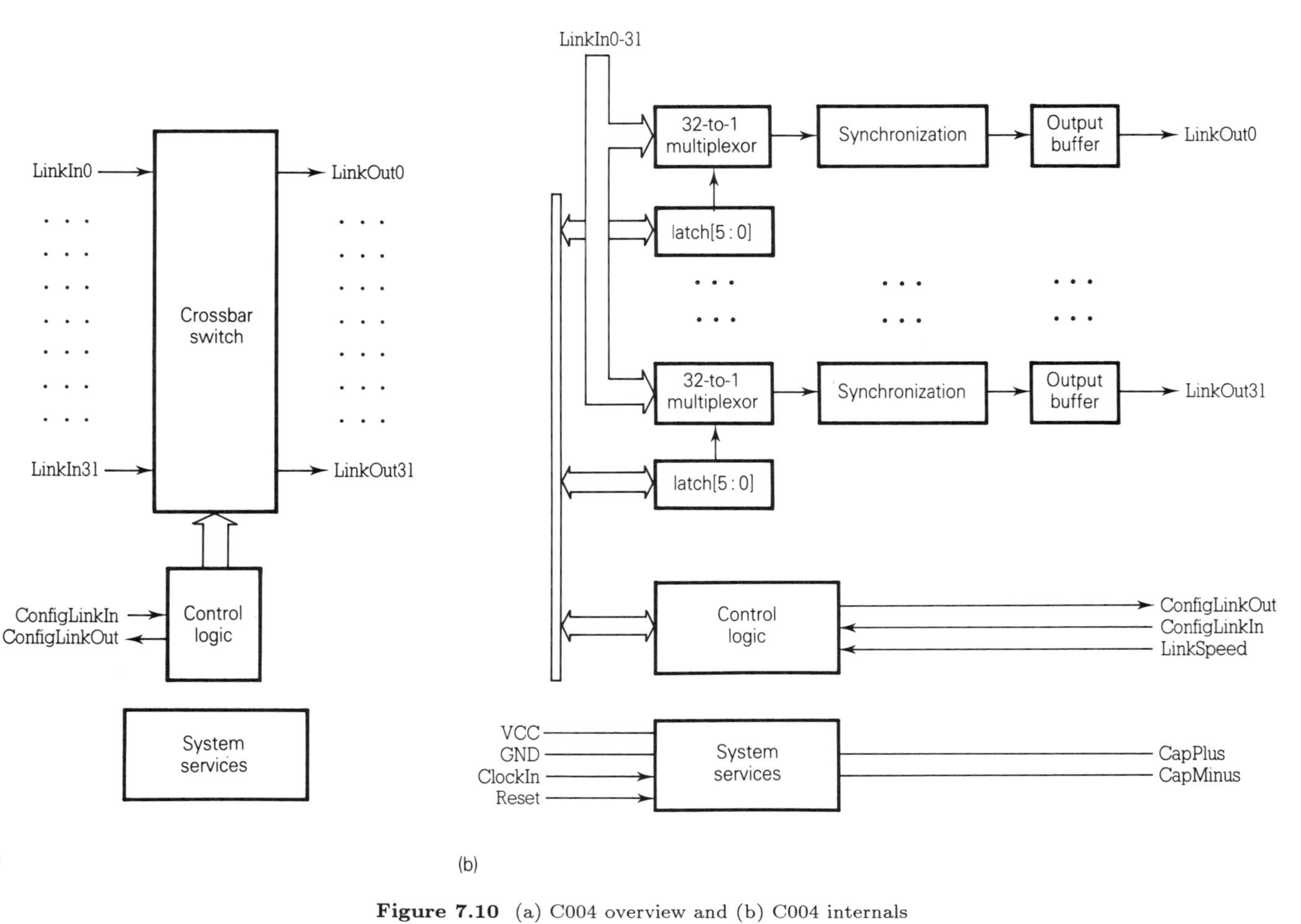

Figure 7.10 (a) C004 overview and (b) C004 internals

The C004 is programmed (see Table 7.6) by byte sequences sent down the configuration link. The reset command removes all connections, and sets all the C004 outputs to the low, inactive, state. The connection commands (0 and 1) load the internal latches with the input number to which the output is to be connected. However, these connections are not made until the setup command is received.

The simple connect command, 0, sets up a connection between one input and one output. For example, to program the C004 so that the only connection is between input 2 and output 6, the following byte sequence is required:

4	reset C004
0 2 6	specify connection
3	activate the C004

The link connection command, byte 1, sets up two connections within the C004. This is useful if the transputer links are wired so that the link inputs and outputs of each link are connected to the same numbered C004 output and input. For example, Figure 7.11 shows link 0 of two transputers, A and B, wired to C004 ports 5 and 10. To connect these links we may either use two simple connect commands:

```
0 5 10
0 10 5
```

or one link connect command:

```
1 5 10
```

The order of the links in the second command string is immaterial, as a bidirectional connection is made.

The state of the C004 connections can be read by the Enquire command. When sent with an output number, this command returns a byte containing the input number to which this output is connected; the top byte of the integer is set if the output is connected, reset if it is disconnected.

The delay between input and output of the C004 is about 1.75 bit times, regardless of whether the links are run at 10 or 20 Mbps. Table 7.7 shows some measurements on data transfer rate made through C004 links.

Table 7.6 C004 programming commands

Command	Function
`0 [input] [output]`	connect `[input]` to `[output]`
`1 [link1] [link2]`	connect `[link1]` input to `[link2]` output and `[link2]` input to `[link1]` output
`2 [output]`	return the input to which output is connected
`3`	sent at the end of every command sequence to activate the device
`4`	reset the device. All outputs are disconnected and held low
`5 [output]`	output `[output]` is disconnected and held low
`6 link1 link2`	disconnects `[link1]` and `[link2]`

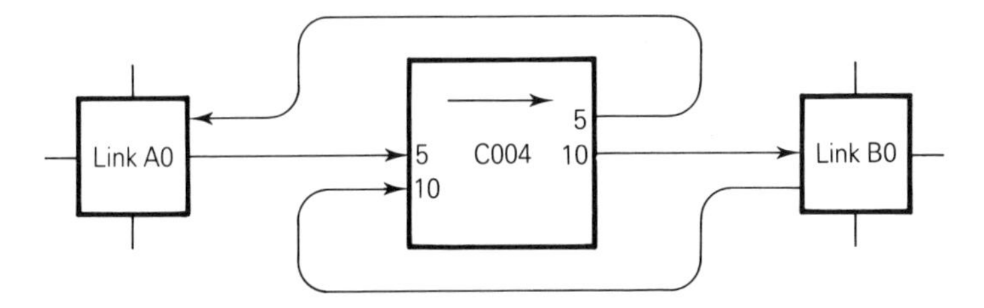

Figure 7.11 Two transputers connected to a C004

Table 7.7 Data rates between transputers connected through various numbers of C004 link switches

Data path	Number of link switches		
	0	1	2
T800 to T800	1.69	1.35	1.07
T414 to T800	1.36	1.06	0.89
T800 to T414	0.91	0.76	0.68
T414 to T414	0.80	0.69	0.60

Measurements made with 20 Mbps links, on 20 MHz T800c and T414b transputers, and shown in units of Mbytes per second

It is clear from these measurements that the link data rate is reduced by about 15–20 percent by a single C004, even for those transputers that have overlapped acknowledge. Adding a second C004 reduces the data rate to about 70 percent of its original value.

Chapter 8

System Integration

The previous chapter described the overall architecture of the transputer range. This chapter is concerned with some of the details of building a system based on transputers.

At first glance this may seem very simple, just a case of adding some memory to various transputers and connecting up the links.

There are, however, a number of points to be considered. For example, the distance between each transputer will affect the link buffering scheme and the maximum data rate available. Different manufacturers have used different types of reset and analyze schemes, and these are rarely compatible.

This chapter does not intend to act as a guide to actually building transputer systems. It does attempt to highlight some of the design decisions required, and various solutions to some of the problems. It is also intended to indicate whether certain products from one manufacturer are compatible with those from another. In order to perform this role the last part of the chapter describes some of the available transputer hardware; this is meant to show the range available at the time of writing (December 1989) and is not to be taken as either an exhaustive list nor as a particular recommendation.

8.1 Connecting transputer links

At the heart of any multi-transputer system lie the links that transfer data from one processor to another (see Figure 8.1). These links operate a very simple byte protocol, where each byte transmitted is acknowledged by the receiving transputer, and there is no hardware error detection or correction. It is important therefore that the link connections should be made with a reliable physical medium, operating well within its specification. For very short distances direct connection is satisfactory, but at longer distances attenuation and reflections affect the magnitude and shape of the received signals. Thus more sophisticated methods have to be used.

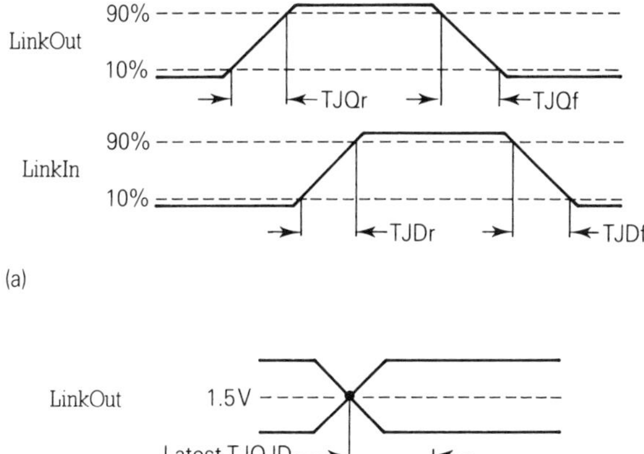

Figure 8.1 Link waveforms

The published specifications affecting the link waveform are shown in Table 8.1; these include typical values for the rise and fall times of the link outputs. There is one additional parameter, the link speed itself. The link bit rate is generated by multiplying up the `ClockIn` signal, and for proper operation of the links the `ClockIn` frequency of two connected transputers should not differ by more than 400 ppm, giving a requirement that each individual clock should be stable to 200 ppm. This is not a very severe requirement, and is easily met by standard crystal oscillators.

Table 8.1 Link waveform specifications

Symbol	Parameter		Min.	Nom.	Max.	Units
TJQr	`LinkOut` rise time			12	20	nsec
TJQf	`LinkOut` fall time			5	10	nsec
TJDr	`LinkIn` rise time				20	nsec
TJDf	`LinkIn` fall time				20	nsec
TJQJD	buffered edge delay		0			nsec
TJBskew	variation in TJQJD	20 Mbps			3	nsec
		10 Mbps			10	nsec
		5 Mbps			30	nsec
CLIZ	`LinkIn` capacitance @ 1 MHz				7	pF
CLL	`LinkOut` load capacitance				50	pF
RM	series resistor for 100 ohm transmission line			50		ohms

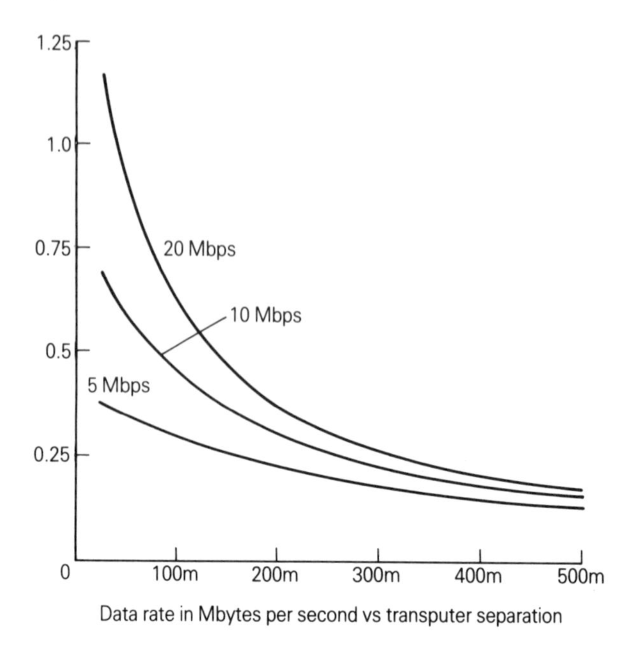

Figure 8.2 Data rates in Mbytes per second vs transputer separation for two T800c transputers

8.1.1 Link data rate

As each byte of a message is individually acknowledged, a delay in the handshake will reduce the data rate of the link. We have seen this when data are sent through C004 link switches (Table 7.7), and the same effect occurs when data are transferred on long physical connections. Figure 8.2 shows how the calculated data rate varies with connection length for the three link speeds. Here we have assumed that two T800c transputers are communicating along a cable with a signal velocity 66 percent of the speed of light. A striking feature of these results is that with increasing distance the advantage of using the higher bit rates becomes almost negligible. This is because the limitation on data rate is the round trip delay, rather than the bit rate. It is much more convenient to use the 5 Mbps bit rate with its less severe skew requirements when designing link connections over 500 yards or so. For very long connections it may be necessary to use a different link technology, acknowledging each message rather than each byte.

8.1.2 Matching

Short links can simply be made by directly connecting the links of two transputers. However, this method breaks down when the connections are sufficiently long for transmission line effects to become important. This occurs when the propagation time along the link is about one-third of the rise or fall time of the signals transmitted along it. The transputer parameters in Table 8.1 show that the shortest transition of `LinkOut` is fall time, which can be as little as 5 nsec for a lightly loaded output. Thus transmission line effects will be noticed for propagation delays of more than about 2 nsec, which corresponds to a length of 16 inches for a cable which has a signal velocity of 66 percent of the velocity of light. This is the origin of the rule of thumb that direct link connections can be made for distances up to 12 inches or so.

Reflections occur on a transmission line wherever there is a discontinuity in its characteristic impedance, and can be reduced by using a constant impedance transmission medium, and by matching the link output to this. This is most simply done by a series matching resistor (Figure 8.3), which gives a more efficient coupling of the link signal into the line, and also reduces the effect on the link output of reflections from the receiver end. The link output can be matched to a 100 ohm transmission line by a resistor of about 56 ohms.

However, some reflections from the receiving end will still reach the link output, and if received during a link signal transition may cause improper operation of the link hardware and loss of data. To avoid this possibility, matched link connections should have a round-trip propagation time of less that about 0.8 bit times, so that

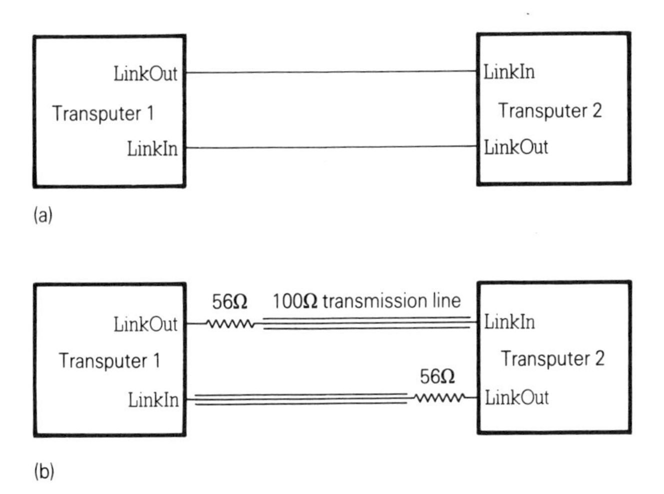

Figure 8.3 Direct and matched link connections (nominal value shown for series matching resistor)

the reflections of each signal edge are received before the next edge is transmitted. This delay time corresponds to a cable length of approximately 3.5 yards at 20 Mbps, 7 yards at 10 Mbps and 14 yards at 5 Mbps.

8.1.3 Link buffering

When longer connections are to be made the link signals must still be transmitted along constant-impedance media, but must also be buffered both to combat attenuation and to avoid the effects of reflected signals on the link output. In designing buffering schemes it is important to meet the rather strict skew requirements of the link signals (Table 8.1). Skew is a distortion of the received waveform caused by changes in the propagation delay of the transmission system. It is often due to variations in the rise and fall times of buffering devices, and can most easily be avoided by using high-bandwidth buffers with short propagation delays.

For short distances links can be buffered by single-ended output devices such as FACT buffers (Figure 8.4a). These have good skew characteristics, but are difficult to match closely to transmission lines. They are also more sensitive to noise than differential drivers and receivers. The use of FACT appears satisfactory up to about 20 yards.

For longer distances differential drivers and receivers are recommended. The RS422 standard (see Figure 8.4b) provides specifications for data rates up to 10 Mbps over twisted pair cable. However the skew specifications of RS422 are not very tight, and the standard does not apply to the 20 Mbps rate. Despite these potential problems the successful use of RS422 devices has been reported at distances of 10 yards or so for 20 Mbps and 30 yards at 10 Mbps.

Differential ECL buffers (Figure 8.4c) provide very high bandwidth with tight skew specifications, and are suitable for 20 Mbps links. Meiko (1987) has proposed a standard for 20 Mbps link connections using the 10H124 quad TTL to ECL buffer and the 10H125 quad ECL to TTL receiver and 100 ohm twisted pair cable. Among the disadvantages of ECL are the additional −5.2V power supply required and the general difficulty of using ECL devices.

8.1.4 Optical fibers

INMOS have described a 5 Mbps fiber optic link that will work at distances of 500 yards or more. From Figure 8.2 it is clear that there is very little advantage in using the higher bit rates at such distances. This is fortunate, as the skew inherent in the optical medium and the devices that drive it becomes far too large for correct operation of links at high bit rates. Table 8.2 shows a summary of link buffering techniques and their characteristics.

Table 8.2 Link connection techniques

Technique	Max. distance	Remarks
Direct	12 in.	cheap
Matched	14 yards	5 Mbps
	7 yards	10 Mbps
	3.5 yards	20 Mbps
FACT buffers	20 yards	potential noise problems as single-ended
RS422	40 yards	at 5 Mbps
	20 yards	at 10 Mbps
ECL buffers	20 yards	at 20 Mbps
Fiber optics	>1000 yards	bit rate limited by skew

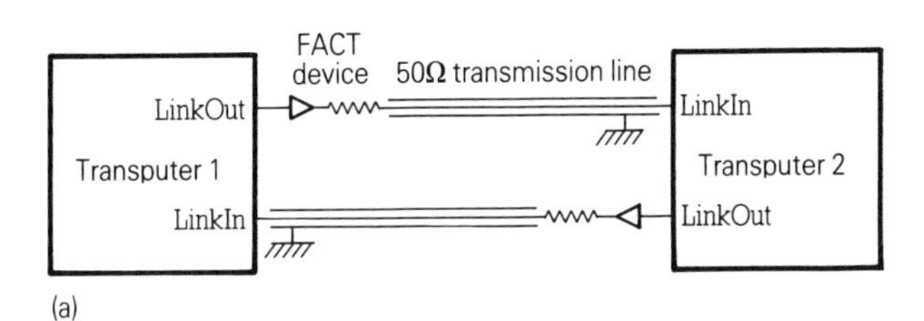

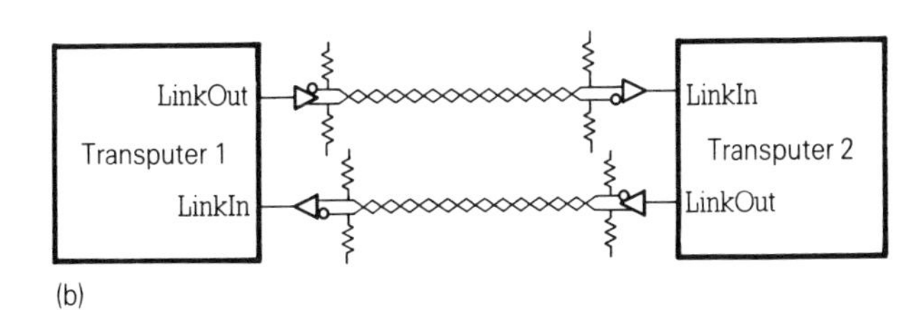

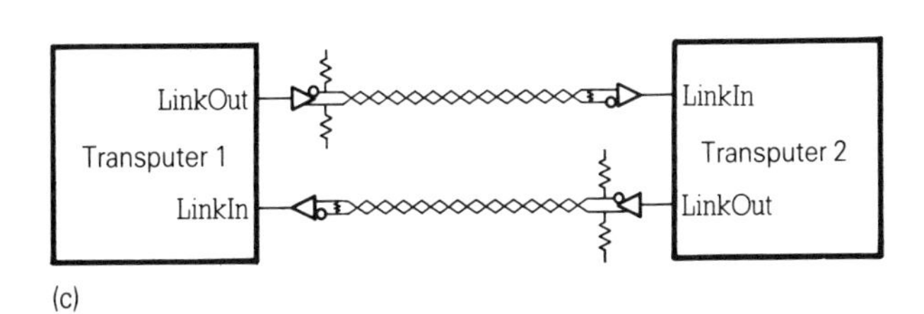

Figure 8.4 Link buffering techniques: (a) FACT buffering with series matching resistor; (b) RS422 drivers and receivers connected by 100Ω twisted pair; (c) ECL drivers and receivers connected by 100Ω twisted pair

8.2 Reset, analyse and error schemes

For a multi-transputer system to be reliable it must be possible to discover when a processor fails, to investigate the causes of that failure, and to restart the offending processor if the error is not too serious. These functions have to be carried out by means of the `Error`, `Analyse` and `Reset` signals, and the links. The best way of connecting these signals in a large system of transputers is not obvious, and several attempts have been made to find a satisfactory solution.

8.2.1 Subsystem reset

In the original INMOS scheme each transputer board has three system services ports: `up`, `down` and `subsystem`, which are used to connect boards together in a hierarchy (Figure 8.5). The `up` port receives `notReset` and `notAnalyse` signals from its ancestor in the hierarchy, and transmits these signals on to its siblings and children through the `down` and `subsystem` ports respectively. The processor on the board can also assert `subsystem notReset` and `notAnalyse`, normally by writing to special locations in its external memory space. These signals are `ORed` with the corresponding signals received at the `up` port. Thus any processor can assert `Reset` or `Analyse` simultaneously for all processors below it in the hierarchy, and the host can reset the whole system.

Error signals are propagated in the opposite direction, `down notError` is `ORed` with the on-board error signal and transmitted through the `up` port. `Subsystem notError` is not automatically propagated, but can be read by the on-board processor, which may then assert its own `Error` output. Thus each processor can be

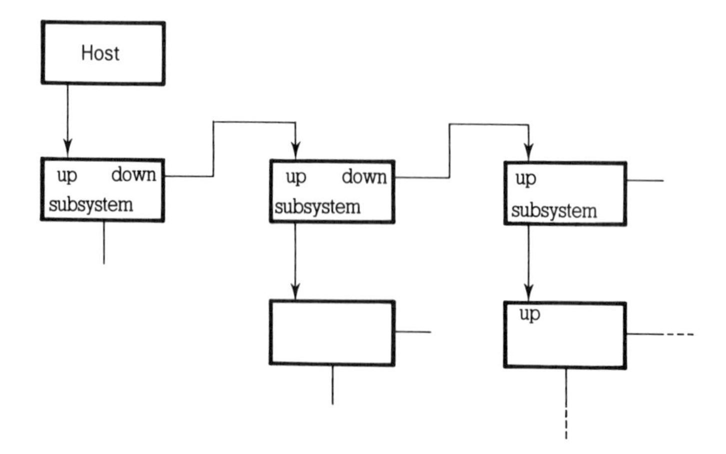

Figure 8.5 Up, down, subsystem

aware of the error state of all of the processors below it in the hierarchy.

It is clear that in this reset scheme it is impossible to reset or examine the error state of an individual processor, except for an isolated processor at the bottom of the hierarchy.

8.2.2 Distributed reset

Parsytec have introduced a distributed reset scheme (see Figure 8.6) where the concept of the link is extended to include bidirectional reset signals. A transputer can reset any transputer to which it is connected by such an extended link. Each transputer has an error latch, mapped into its external memory space, which serves to store the origin of any error. Three kinds of error can be detected: internal errors that assert the transputer's `Error` output, memory parity errors, or address errors caused by the transputer attempting to access non-existent memory. When any error occurs `Analyse` is asserted, which causes the offending transputer to halt.

A processor connected to the halted transputer can now reset it, and use peek operations along the link to determine the cause of the error. It may then attempt to reboot the transputer over the link.

8.2.3 Bus systems

One of the earliest large-scale commercial transputer systems was the Meiko Computing Surface. The architecture of this product is outlined in Section 8.6.1; here we are concerned only with the reset scheme. Each transputer board in the Computing Surface is connected by a proprietary interface device to the Supervisor Bus. This is an 8-bit parallel bus, and is controlled by a transputer board known as the

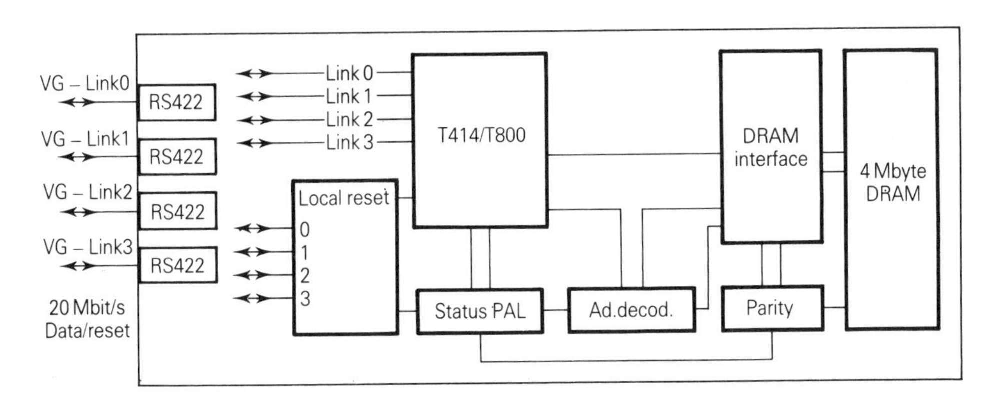

Figure 8.6 Distributed reset

Local Host. The Supervisor Bus has several functions:

- It acts as a route for debugging messages.
- It supports software and hardware for failure detection and reporting.
- It controls electronic configuration of the inter-processor links.
- It provides a reset and analyze function for individual transputers or for the whole or sections of the machine.

This clearly provides a very general reset and analyze mechanism, but at the cost of quite a complex bus interface at each processor.

The Atari Transputer Workstation also uses a bus system for reset control. However, this is a very simple system with only two wires called `fast` and `slow`, which are daisy-chained through the processors. Its position on the daisy-chain assigns each transputer a small integer address. When a transputer is to be reset, `slow` is asserted and `fast` pulsed a number of times. A circuit attached to each transputer counts the number of times that `fast` is pulsed before `slow` is deasserted, and resets the transputer if this number corresponds to its address. The protocol supports the assertion of `Reset` and `Analyse`, either together or separately, for a single processor or for all processors together. The whole of this logic can be implemented in a single PAL.

8.3 Transputer modules and motherboards

Although the transputer is a useful component in its own right, there are certain disadvantages in constructing equipment with 'raw' transputers. For one thing, the amount of on-chip memory is insufficient for the majority of applications. The transputer must be mounted on a suitable printed circuit board (PCB), power and ground supplied at low impedance with sufficient decoupling, and the memory interface designed. To avoid many of these problems the transputer module or TRAM was developed. A TRAM consists of one or more transputers mounted on a PCB, with memory and perhaps other interface circuitry (see Figure 8.7). The TRAM in turn is mounted on a motherboard that supplies power and distributes the clock; many motherboards also have one or more C004 link switches for reconfiguration of the TRAM-mounted transputer links. INMOS and other companies now supply a wide range of TRAMs and motherboards, and various standards have been defined.

8.3.1 Module architecture

The basic size of a TRAM is 1.05×3.66 inches; this is referred to as size 1. Larger and smaller TRAMs (Table 8.3) can be constructed based on this basic module, most commonly by increasing the overall width. Sixteen connections are made to the TRAM in two groups of eight, one at each end of the module.

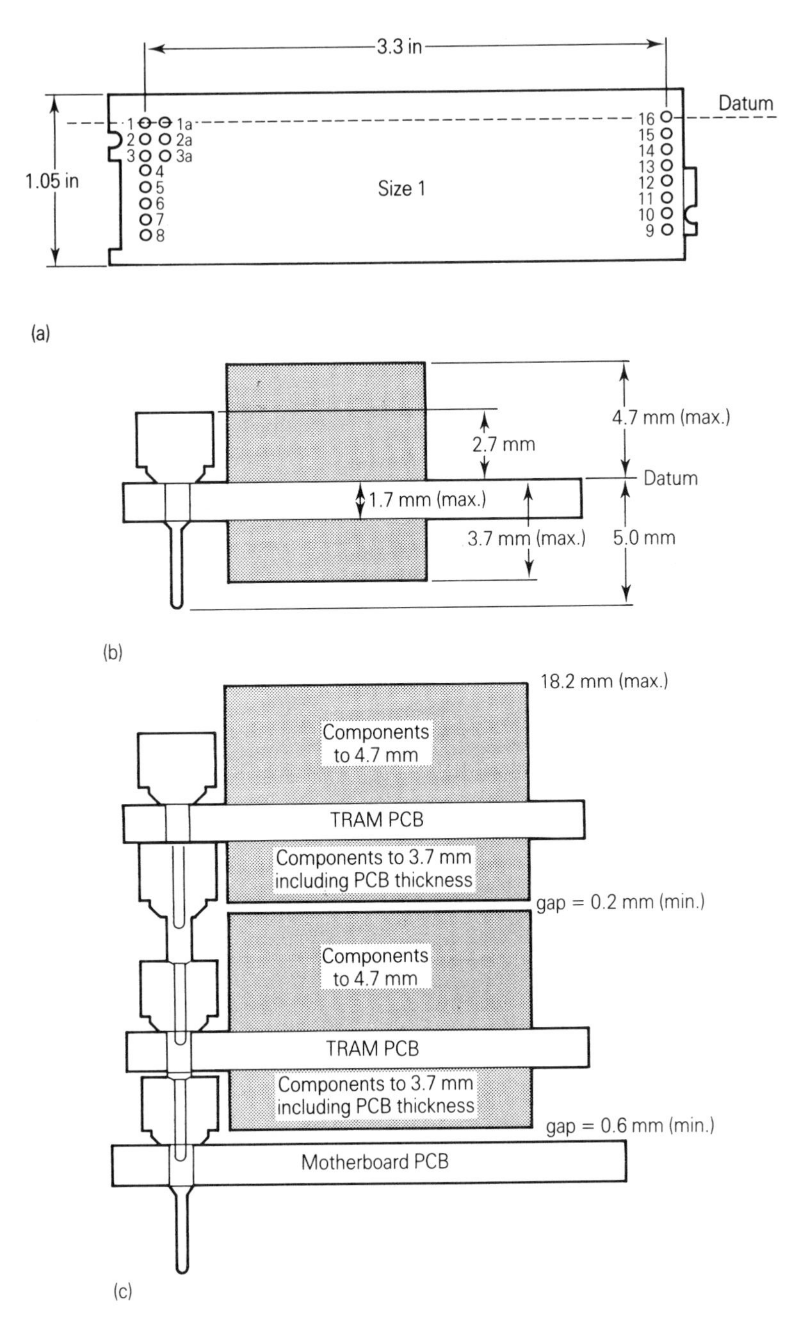

Figure 8.7 TRAM geometry

Table 8.3 TRAM sizes

Size	Distance between pin rows (in.)	Width overall (in.)
1/4	0.6	1.05
2/4	0.6	2.15
1/2	1.5	1.05
1	3.66	1.05
2	3.66	2.15
4	3.66	4.35
8	3.66	8.75

The signals brought to the pins of a TRAM are kept to a minimum: the power pins `Vcc` and `GND`; `Reset`, `Analyse` and `notError`; four links; and `LinkSpeedA` and `LinkSpeedB` to select the speed of the links. If both `LinkSpeedA` and `LinkSpeedB` are low the links will operate at 10 Mbps, if they are both high the links operate at 20 Mbps. Other combinations are reserved for future enhancement, and it is assumed that 5 Mbps will not need to be used. The `Error` pin of the transputer is inverted at the `notError` TRAM output, and driven by a open collector or open drain circuit. Thus the `notError` signals of several modules may be wire `ORed` together. This means that the `ErrorIn` pin is not used on transputers mounted on modules, and should be held to ground. Table 8.4 gives the TRAM pinouts.

It may be useful for a TRAM to control a group of other transputers or TRAMs. Conventionally this group is known as a subsystem, and three subsystem control signals are defined: `SubSystem Reset`, `SubSystem Analyse` and `SubSystem notError`. Pin positions have been allocated in the TRAM specification for these signals, positioned inside pins 1, 2 and 3 of the standard module (Table 8.5).

The subsystem is controlled by reading and writing at memory locations above memory address zero. A module may have multiple subsystem outputs, and if so

Table 8.4 TRAM pinouts

Link2Out	1	16	Link3In
Link2IN	2	15	Link3Out
Vcc	3	14	GND
Link1Out	4	13	Link0In
Link1In	5	12	Link0Out
LinkSpeedA	6	11	notError
LinkSpeedB	7	10	Reset
ClockIn	8	9	Analyse

Table 8.5 TRAM subsystem pinouts

Pin	Signal
1a	SubSystem notError
2a	SubSystem Reset
3a	SubSystem Analyse

Table 8.6 TRAM memory map for 32-bit transputers

Address	Description
`0x7FFFFFFF`	top of address space
`0x7FFFFFFE`	ROM bootstrap
}	peripheral space
`0x0000000C`	
`0x00000008`	parity control
`0x00000004`	`SubSystem Analyse`
`0x00000000`	`SubSystem Reset`
`0x8FFFFFFF`	
}	user RAM, internal and external
`MemStart`	
}	reserved internal RAM space
`0x80000000`	

their control locations are separated by 16 bytes. The base address is `0x00000000` for 32-bit transputers, and a write into this location sets `SubSystem Reset` to reflect the state of bit 0 of the byte written. A read from this location returns the value of `SubSystem Error` in bit 0. Location `0x00000004` is used for `SubSystem Analyse`, a write here asserts this signal if bit 0 is 1, deasserts it if bit 0 is 0 (see Table 8.6). Note that these locations can often be written to accidentally in a language such as C, where 0 is used as a null pointer value. The effect may be to reset *another* transputer, leading to a very subtle bug.

TRAMs may implement memory parity checks. If so, the parity logic is controlled through a register at address `0x00000008` on 32-bit transputers. Writing a 1 to bit 0 of this register should enable parity checking, writing a 0 should disable it. Reading the byte returns the status of the parity check. Bit 0 is set if a parity error has occurred, bits 1 and 2 indicate in which byte it has occurred, and the other bits may contain information showing which memory bank is in error. It is up to the designer and programmer to decide what action should be taken when a parity error occurs, but normally `notError` will be asserted, and data transfers from the offending transputer halted.

8.3.2 Typical TRAMs

A size 1 TRAM can contain a transputer and about eight memory devices. At the present level of technology, this corresponds to 1 Mbyte of dynamic memory, or around 256 Kbytes of static RAM. However, as memory densities increase, the amount of memory available on a size 1 TRAM will increase to 4 or 16 Mbytes.

The present INMOS TRAMs are shown in Table 8.7. Several third-party manufacturers also produce TRAMs, and although some manufacturers compete in the compute-only market, most have preferred to produce special-function TRAMs, such as graphics and peripheral interfaces. A list of TRAM manufacturers is given in Appendix C. It is now possible to construct a complete transputer workstation using only TRAMs and a suitable motherboard.

8.3.3 Motherboards

The minimum function of the motherboard is to provide power and clock distribution, to make link connections between modules and to `OR` together the `notError` signals from the modules into a board `Error` signal. Additionally, many motherboard designs have C004 link switches on board, and provide external interfaces, for example to the IBM PC or VME buses.

INMOS has described a standard link configuration for motherboards, shown in Figure 8.8a. This has the module sites connected in a pipeline arrangement, with link 2 of each module wired to link 1 of the next. Link 1 of the first module and

Table 8.7 INMOS TRAMs

Compute only

Part no.	Transputer	Memory size/cycles	Subsystem	Size
B401	T414-20, T425-25 or T800-25	32K/3 SRAM	no	1
B402	T222-20	8K/2 SRAM	no	1
B403	T414-20, T425-20, or T800-20	1M/3 DRAM	yes	4
B404	T414, T425 or T800	32K/3 SRAM, 2M/4 DRAM	yes	2
B405	T800-20	8M/5 DRAM	yes	8
B410	T801	160K/2 SRAM	no	2
B411	T425-20 or T800-20	1M/3 DRAM	no	1
B416	T222	64K/2 SRAM	no	1
B417	T800	64K/3 SRAM, 4M/4 DRAM	yes	4

Special application

Part no.	Application	Description
B408	graphics	T800-20, 1M/4 DRAM, 1.25M/4 dual port RAM (use with B409), max. resolution 1024 × 768, size 8
B409	graphics	T222-20, video timing generator, color lookup tables (use with B408), max. dot rate 64 MHz, size 8.
B419	graphics	T800-20, G300 color video controller, 2M/4 DRAM, 2M/4 VRAM, max. video resolution 1280 × 1024, 8 bits/pixel, size 2
B407	ethernet	T222-20, 64K/3 SRAM Am7990 lance
B422	SCSI	T222-20, 64K/2 SRAM, subsystem, target/initiator, size 2
B418	ROM	T222-20, 256K flash ROM, subsytem, size 2
B421	GPIB	T222-20, 48K SRAM, 8K EEPROM, GPIB controller, size 4
B415	link	RS422 buffer for 4 links, reset and system services, size 1
B420	vector	vector processing TRAM. T800-25, 1M DRAM, 256K dual-port SRAM, vector processor, size 2

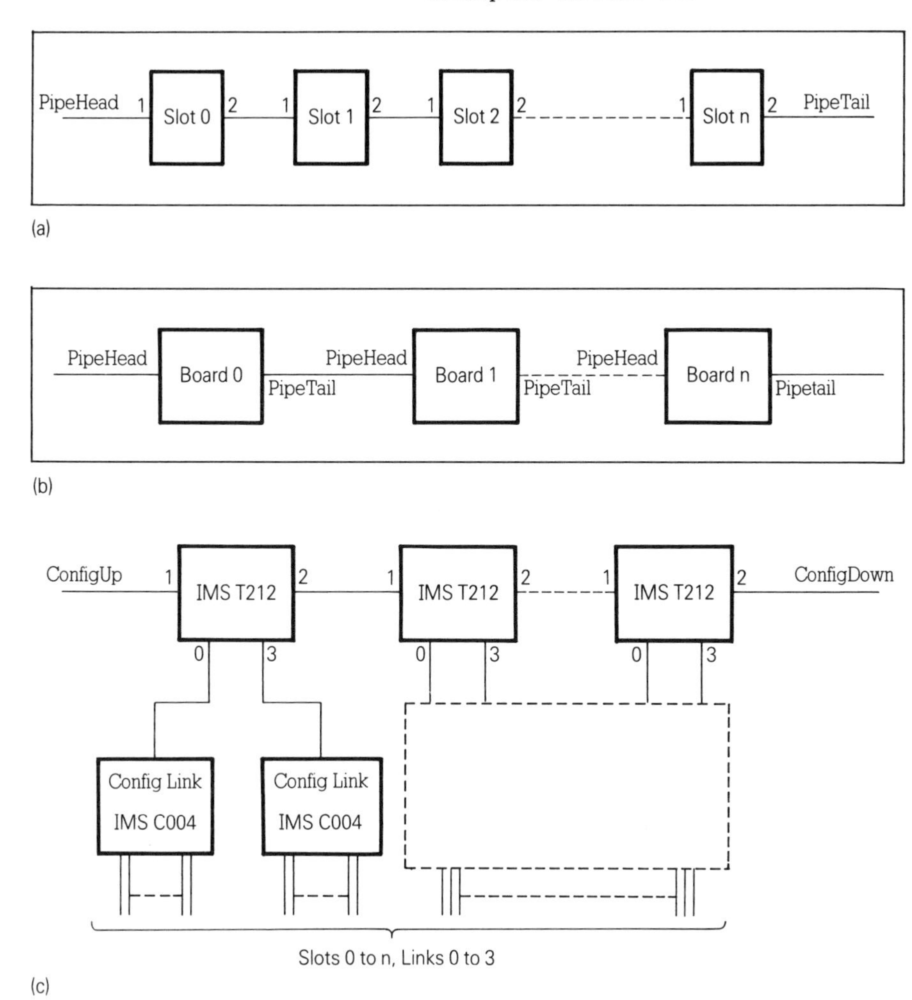

Figure 8.8 Motherboard architecture

link 2 of the last are taken off the board as the `PipeHead` and `PipeTail` respectively. Thus several boards can be connected together to form a longer pipeline (Figure 8.8b). Links 0 and 3 of each module are taken to a C004 link switch, of which the board may have more than one. Each C004 is controlled by a T212 (or T222) transputer, and these configuration transputers are also linked into a pipeline from board to board (Figure 8.8c).

Figure 8.9 shows the complete block diagram of the B008 IBM PC motherboard. This has ten module sites, arranged in the standard pipeline through links 1 and 2. Links 0 and 3 of each module are taken to a single C004 link switch, so that any transputer can be connected to any other through two links. The B008 is

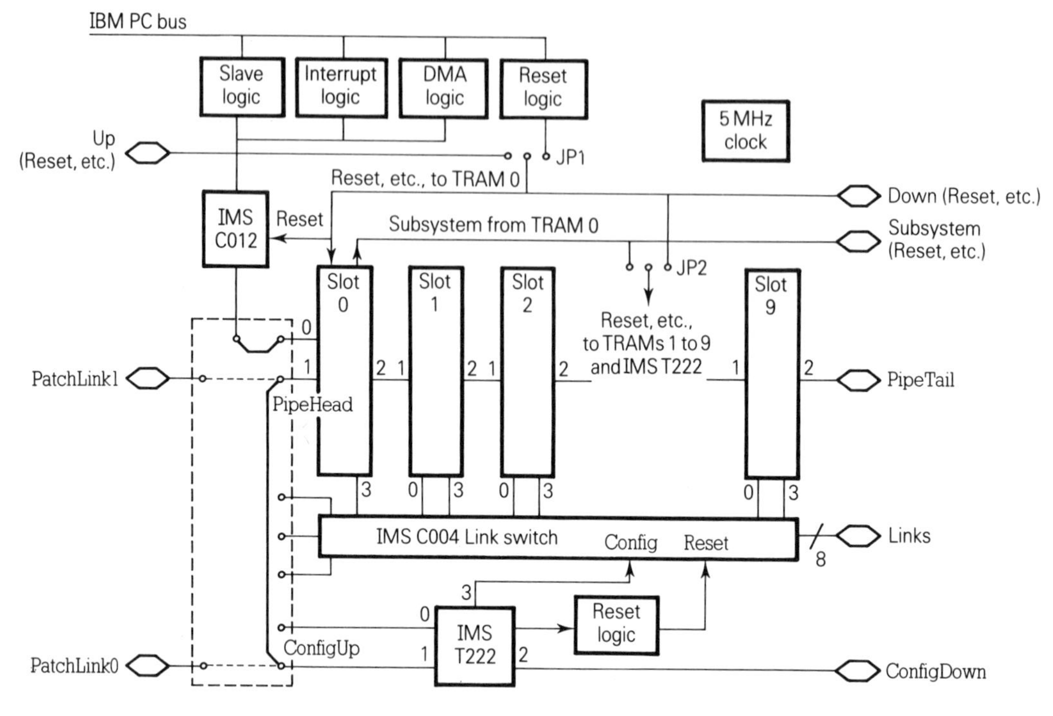

Figure 8.9 B008 block diagram

interfaced to the PC bus with a C012 link adaptor. This link can be patched into the PipeHead, into module 0 link 0, or into any other module via the link switch. The link switch itself can be reset and configured by the on-board T222 controller.

The B012 double eurocard motherboard has a rather different link interconnection scheme (see Figure 8.10). This board has sixteen module sites, arranged in the standard pipeline. There are two C004 link switches with the link 3 outputs and the link 0 inputs connected to one C004, and the link 3 inputs and the link 0 outputs connected to the other. Thus link 3 on any transputer can be connected to link 0 on the same or any other transputer, the connection going once through each of the C004s. Clearly, relevant connections of the C004s must be configured identically for this to work. The link delay between two transputers is not increased by using the two C004 devices, as the link signals go through each only once. Spare C004 inputs and outputs are taken to the edge connector to make interboard connections.

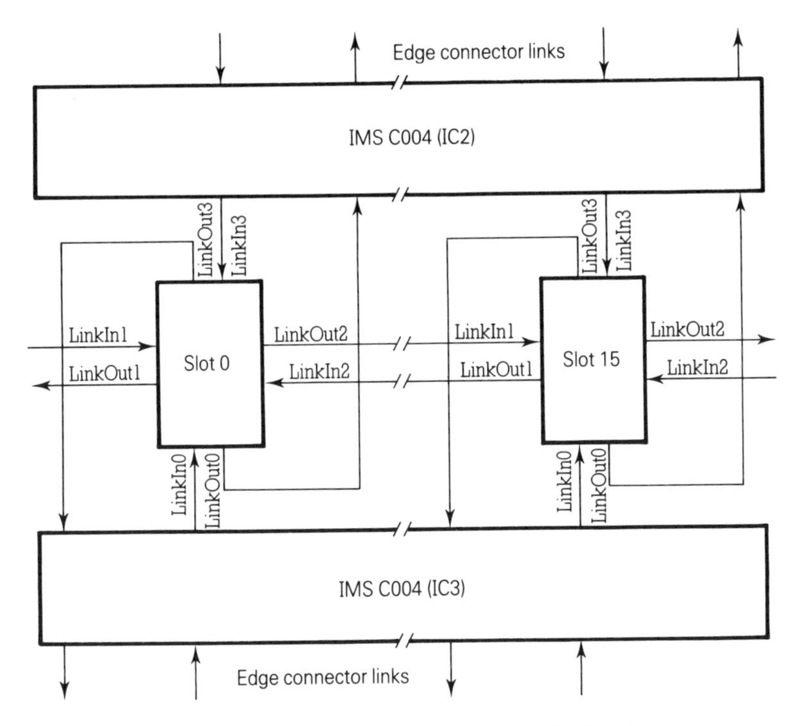

Figure 8.10 Link interconnections on the B012 motherboard

8.4 PC and other boards

Shortly after INMOS produced the first transputer they announced a range of board-level products that they hoped would enable transputer users to get started quickly on application development. Initially four board types were produced: the B001 with a T414 and 32 Kbytes of static memory; the B002, a T414 with 2 Mbytes of dynamic RAM and two RS232 ports; the B003, four T414 each with 256 Kbytes of memory; and the B004, a PC bus board with a T414 and 2 Mbytes of dynamic RAM. This last proved immensely popular as it was able to run the Transputer Development System in a low-cost environment. The design was published and this led to a large number of imitations and compatible products produced by third-party manufacturers.

Since then the range of transputer boards has grown, until now there are more than one hundred different types available. In Appendix C we have attempted to list some of the presently available transputer products, both self-contained boards and TRAMs. This list will of course be found to be incomplete, but should act as a rough guide to what is available and from whom.

8.4.1 PC boards

As mentioned above, the B004 was the original PC board. It had a slow T414-15 processor and was interfaced to the PC bus with C001 link adaptor (now an obsolete device). The PC bus interface was quite slow, and did not use DMA, but it was adequate to run the TDS. INMOS have since replaced the B004 with the B008 TRAM motherboard (Figure 8.9).

Many new boards have now been produced, compatible with the B004 standard but with additional features such as faster processors, additional memory and DMA bus interfaces. With the present level of technology it is possible to put onto a single PC board a single transputer with up to 16 Mbytes of memory, or up to seventeen transputers, including one with sufficient memory to run the TDS or a compiler. This represents an enormous scaling up of the processing power of the simple PC, and various software products have been brought out to take advantage of this hardware. A good example is the Racal-Redac Bloodhound autorouter, which is sold as an add-on to their Cadstar PC-based PCB design system. The package includes the autorouter software, the Helios operating system and a T800 PC card.

The other area of great growth has been in graphics boards. The T800 is a reasonably good graphics processor in itself, and has also been combined with a high-speed graphics processor in PC boards.

8.4.2 Other boards

As well as PC products a large range of transputer boards has been developed to other bus standards, and for standalone transputer systems. For those dissatisfied with the PC environment several manufacturers have produced boards for the Sun range of workstations, as well as VME bus boards that can be used in Suns or other VME systems. Most of these boards have TRAM sites, and so can be populated with INMOS or other manufacturers' TRAMs.

An interesting range of boards is produced by Parsytec. Their Megaframe system consists of a number of products designed for industrial control applications. There are processor boards with up to 4 Mbytes of memory, interface, I/O and video boards. These are interconnected by a common backplane that carries link and reset signals, using the distributed reset scheme described in Section 8.2.3. However, each of these boards can also be used in one of a number of standard backplanes, e.g. VME, PC, Siemens SMP and Kontron ECB, by means of bus bridgehead modules that sit between the board and the backplane.

8.5 Workstations

The PC was the original transputer workstation, and perhaps the majority of transputer development systems are now PC based, running the TDS or an operating system such as Helios. However, there have been some attempts to build completely new workstations based on transputers, but perhaps using other processors for I/O operations.

8.5.1 The Atari Transputer Workstation

The Atari Transputer Workstation (ATW) was first produced in 1988. The heart of the system is a T800 with 4 Mbytes of memory, 1 Mbyte of dual-ported video memory, and a bit blitter. The system memory is expandable up to 16 Mbytes. The T800 is connected by one link to a Motorola 68000 I/O processor, which provides access to keyboard, mouse and disk. The other three links can be connected to 'farmcards', containing four T800 transputers each with 1 Mbyte of four-cycle memory. Up to three farmcards can be accommodated in the standard box, with an electronic link configuration board containing two C004 link switches. Expansion boxes are available for larger systems with further electronic link configuration modules. The ATW runs Helios and X Windows as the standard user interface.

8.5.2 Cogent Research

The US company Cogent Research has produced a transputer workstation series called the XTM. Each workstation contains two T800 transputers, a graphics display and a keyboard. Connections are provided via buffered transputer links to a central node. This contains up to sixteen transputers and a hard disk, along with Nubus connections for expansion boards.

The system software is written in Linda, implemented on top of a cut-down version of UNIX called QIX. The graphics interface is driven via Display PostScript and a version of the News windowing system.

8.6 Large-scale transputer systems

Although INMOS originally designed the transputer for process control purposes it has proved a powerful building block for designers of supercomputer systems using multiprocessor architectures. The built-in links make the design of such large-scale computers reasonably simple, although large systems bring their own set of problems. Meiko was the first company to produce large-scale commercial transputer systems.

8.6.1 The Meiko Computing Surface

A Computing Surface (Chesney and Ganz, 1989) consists of a number of computing elements connected by the link network and a supervisor bus. Each element (Figure 8.11) has one or more transputer processors, memory, link and supervisor bus interfaces, and perhaps some element-specific hardware. The elements are controlled by the Local Host, a transputer processor which is the master of the supervisor bus, controls the link network and interfaces the system to the host processor.

A recent product is the In-Sun Computing Surface, which has all the components of a computing surface on a single Sun board.

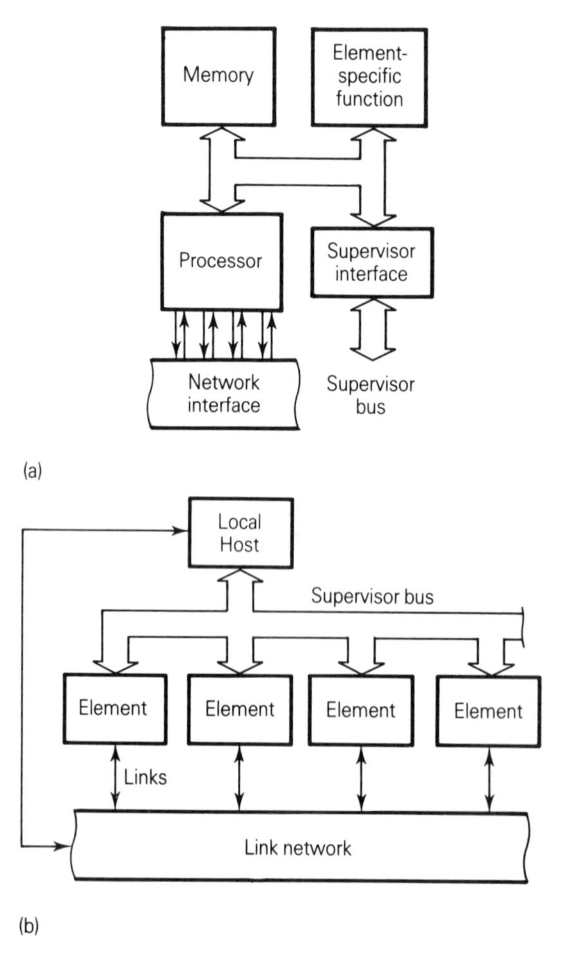

Figure 8.11 Computing Surface element architecture

8.6.2 The Supernode

A European Community ESPRIT I project in the mid-1980s provided some of the funding for the development of the T800 transputer and also supported a consortium of universities and commercial organizations in the development of the 'Supernode', a seventeen transputer building block for supercomputers (Harp, 1987; Harp *et al*, 1987). Two commercial products emerged from this project, the Telmat T-node and the Parsys SN series systems.

The Supernode (see Figure 8.12) contains sixteen worker transputers with a control transputer. All links from all processors are connected to a programmable electronic switch. Each transputer is connected into the proprietary control bus system, implemented as a gate array. This bus has a master–slave protocol with the control transputer as master of the bus. It can reset, analyze, examine the error state and reboot each transputer.

The hardware modules consist of worker boards of eight transputers with various memory sizes, e.g. 256 Kbytes static, 1, 2 and 4 Mbytes dynamic. There are memory server modules, consisting of a T414 with 16 Mbytes, and disk servers with a T414, 16 Mbytes memory and SCSI and floppy disk interfaces.

8.6.3 Parsytec Supercluster series

The Supercluster has a hierarchical cluster-oriented architecture with the T800 or T801 processing element at the lowest level. Sixteen processing elements and

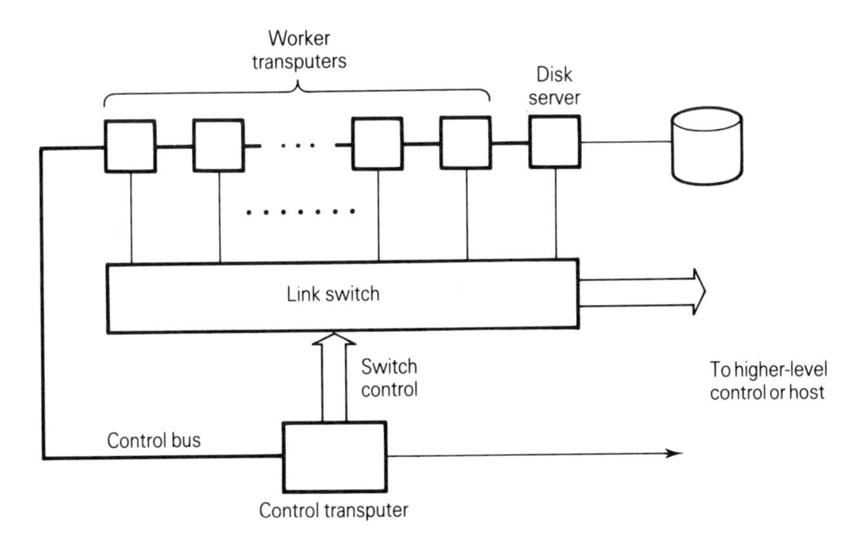

Figure 8.12 Supernode architecture

a network configuration unit are combined to form a computing cluster (Figure 8.13). Four computing clusters form the smallest Supercluster, which also has a System Services Cluster containing mass storage controllers, and user interface and applications-specific devices; see Figure 8.14. The Supercluster has two additional network control units for communication between the computing clusters. Users' workstations can also be connected into the system.

The network control unit (Figure 8.15) consists of T212 and T414 control processors which drive two 96 × 96 configuration matrix switches. Using the Parsytec distributed reset scheme, each link has both data and reset signals, and so one matrix switch configures the data line while the other configures the reset lines. The data switches use C004 link switches, whilst a proprietary device is used for reset signal switching.

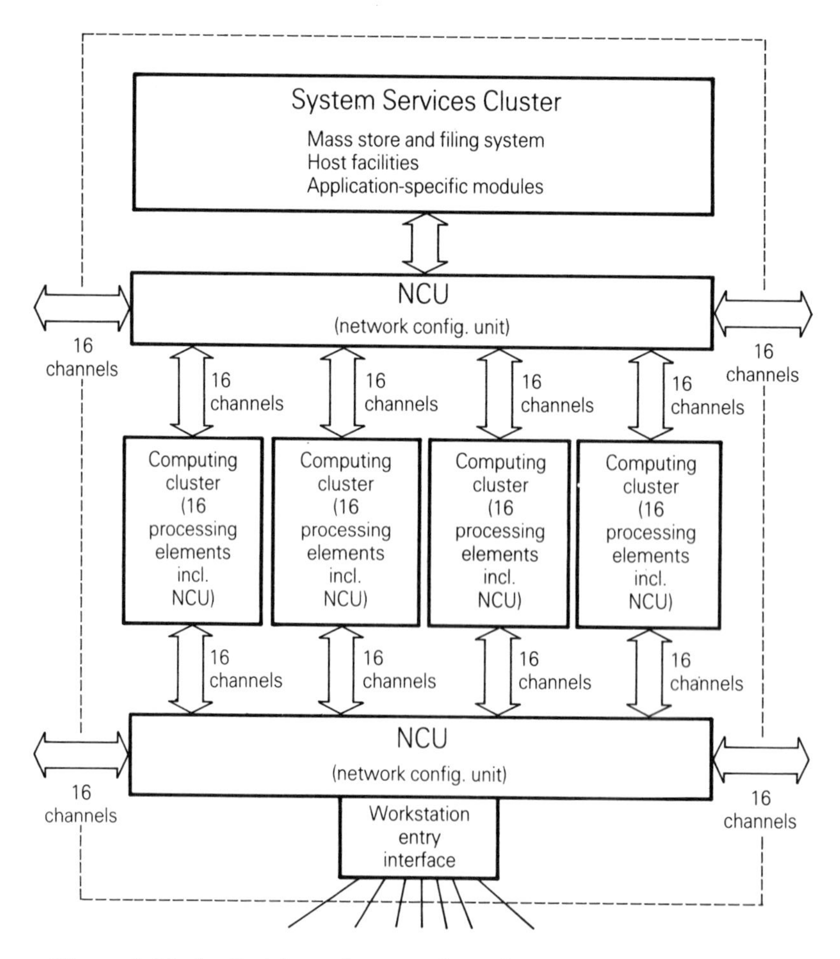

Figure 8.13 Smallest Supercluster configuration

The mass storage controllers consist of a T414 processor with 4 Mbytes of memory and a SCSI interface. There can be up to four of these controllers in each system services cluster, daisy-chained through communications links. Supercluster machines are compatible with the Megaframe series of boards described above,

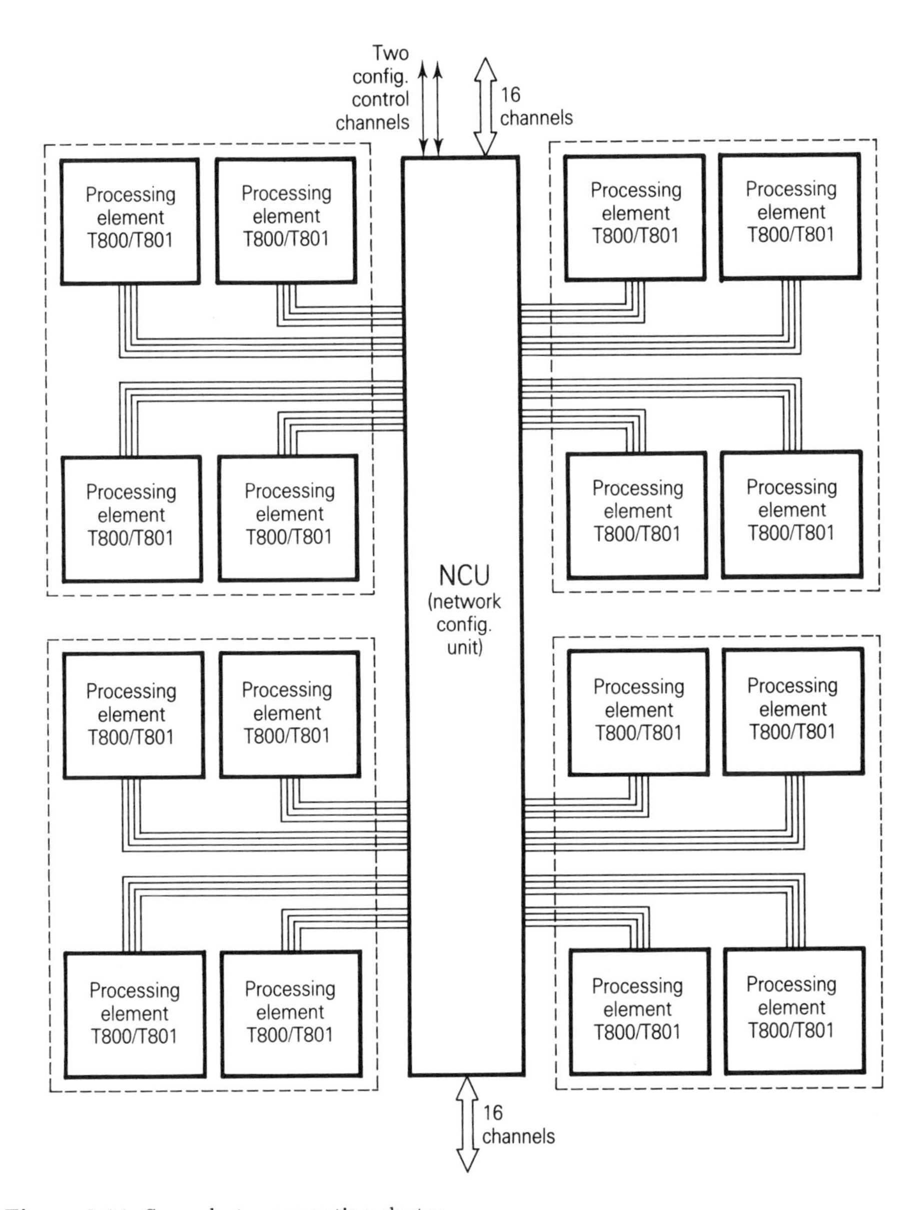

Figure 8.14 Supercluster computing cluster

which provide applications-specific hardware and also access to other buses through bridgehead devices.

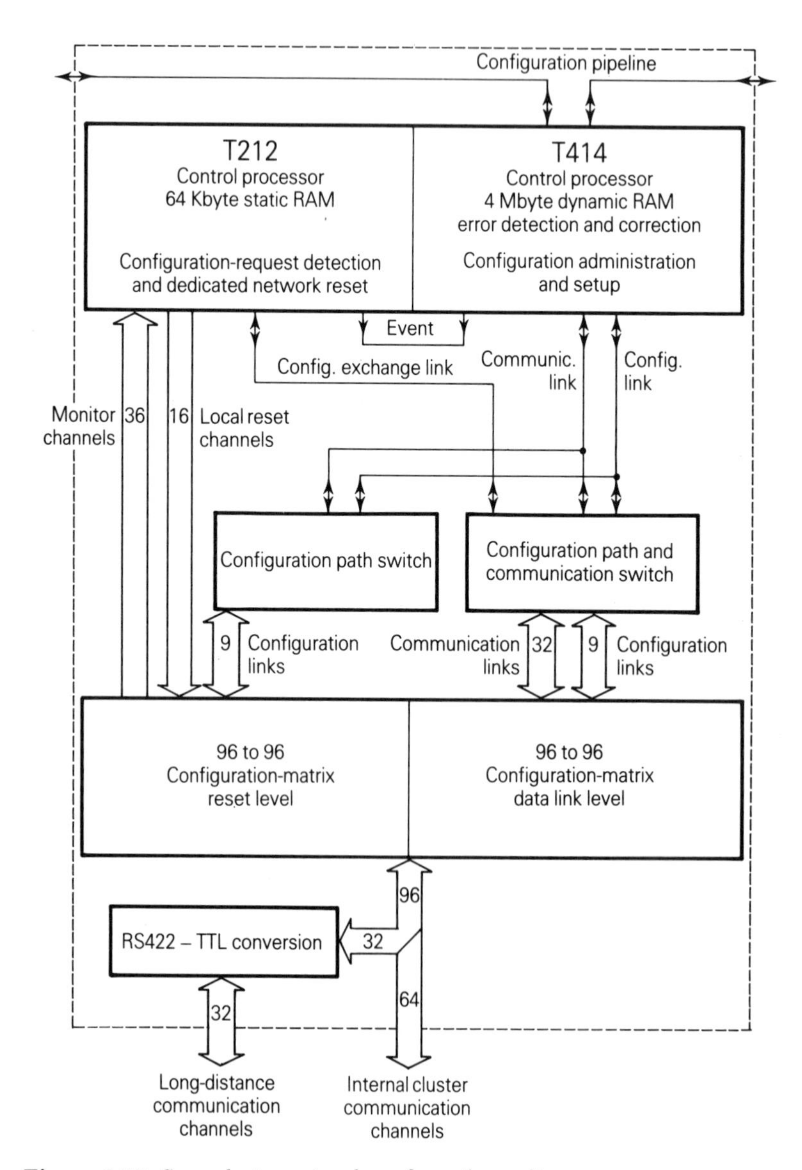

Figure 8.15 Supercluster network configuration unit

Chapter 9

Transputer Hardware Design

This final chapter provides a further level of detail in the hardware design of transputer systems. We describe some of the different options available when adding external memory to both 16-bit and 32-bit transputers. The remainder of this chapter describes a design for a complete size 2 TRAM.

9.1 Sixteen-bit transputer memory interface

The T212, T222 and M212 have a conventional 16-bit memory interface optimized for static memory access. The interface has separate address and data buses, and two byte-write strobes which specify which byte is to be modified in write operations. A chip enable signal selects external memory for both read and write operations, and a wait input can be used to lengthen the memory cycle. All timing of the interface is referred to the processor clock output `ProcClockOut`. Table 9.1 summarizes the 16-bit interface signals.

One unusual feature of the interface is an input signal `MemBAcc`, which can be used to indicate that byte-wide memory is connected. If this signal is asserted at the beginning of a memory cycle, the cycle is split into two byte read or write cycles,

Table 9.1 16-bit memory interface signals

Pin	I/O	Function
`MemA0-15`	out	16-bit external memory address
`MemD0-15`	in/out	16-bit bidirectional data bus
`notMemWrB0-1`	out	byte-write strobes
`notMemCE`	out	chip enable for memory
`ProcClockOut`	out	processor clock output
`MemBAcc`	in	specifies byte-wide memory
`MemWait`	in	delays the memory cycle

using A0 to indicate which byte is to be accessed. If all of the external memory is byte-wide, then MemBAcc can be wired permanently high.

The normal memory cycle takes two cycles of the processor clock, and is divided into four Tstates, T1 to T4 (Figure 9.1), each half a period of ProcClockOut long.

At the beginning of a read cycle, address data are placed on the address bus, are valid at the start of T1, and remain valid until the end of T4. During T1, notMemWrB0 and notMemWrB1 go high (inactive), followed by notMemCE going low (active). The memory data must now be put onto the bus, at a specified setup time before the data are latched during T3. Data must be removed from the bus by the end of T4. If the memory devices cannot meet the data setup requirement, T2 can be extended by taking MemWait high within a short time of notMemCE becoming active. There is also a minimum hold time for MemWait after notMemCE goes active.

The address bus has the same timing in a write cycle as for the read. Shortly after the start of T1, one or both of the byte-write strobes will go low, to indicate which bytes must be written to. Then notMemCe will go active, and the write data

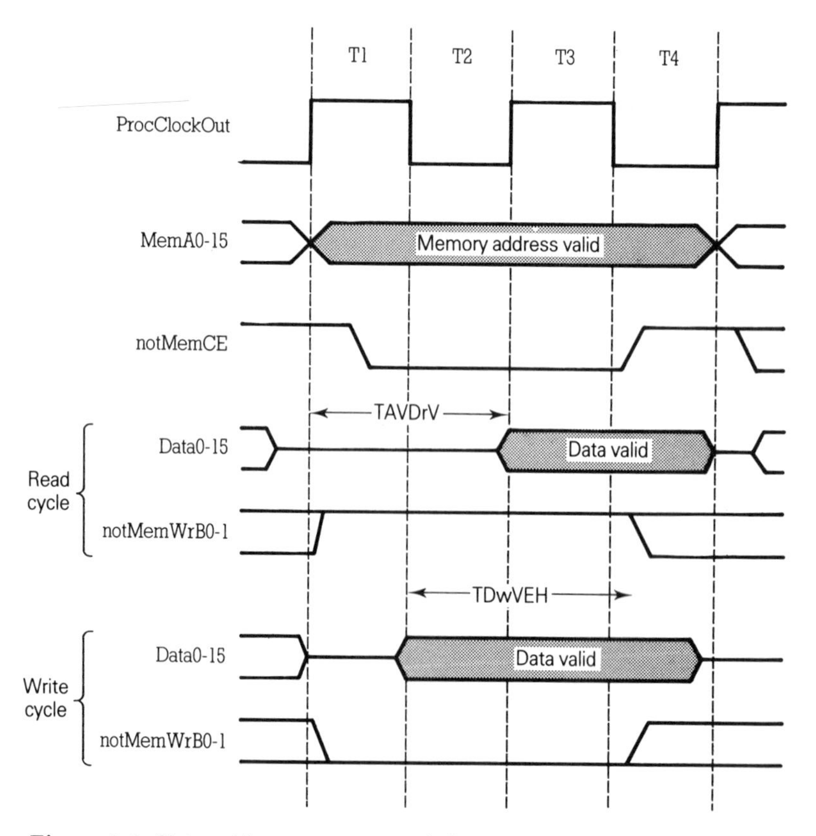

Figure 9.1 Sixteen-bit memory access timing

will be placed on the bus at the start of T2, and removed at the end of T4. These data should be latched into memory on the rising end of notMemCE, which occurs near the end of T3. Again, if the data setup time is insufficient for the memory used, T2 can be extended by the MemWait signal. If the MemBAcc pin is taken high at the beginning of any word read or write cycle, then that cycle will be split up into two byte-access cycles. If all of the external memory is byte-wide then MemBAcc can be wired permanently high.

A minimum parts-count memory layout is shown in Figure 9.2 using two 32K× 8-bit static RAM devices. This design provides 60 Kbytes of external memory on a T222; the only chip decoding required is to use the byte strobes to select the device to be written to on byte-write operations. The T222 memory timing parameters are shown in Table 9.2. The critical parameter for a read cycle is TAVDrV, the time that the memory address is valid before read data must be available. On a write cycle the critical parameter is TDwVEH, the data setup time before chip enable goes high.

Eight-bit wide static devices such as the Toshiba TC55257 are not normally available with access times less than about 85 nsec; in Figure 9.2 the wait state generator lengthens the memory cycle to three processor clock cycles. The D-type flip-flop holds MemWait high until the rising edge of ProcClockOut at the end of T2. As MemWait is sampled during T2, wait states W1 and W2 are inserted into the memory cycle. MemWait is taken high again at the rising edge of ProcClockOut after notMemCE goes high, i.e. at the end of T4 (see Figure 9.3). Lower density 16K×4-bit

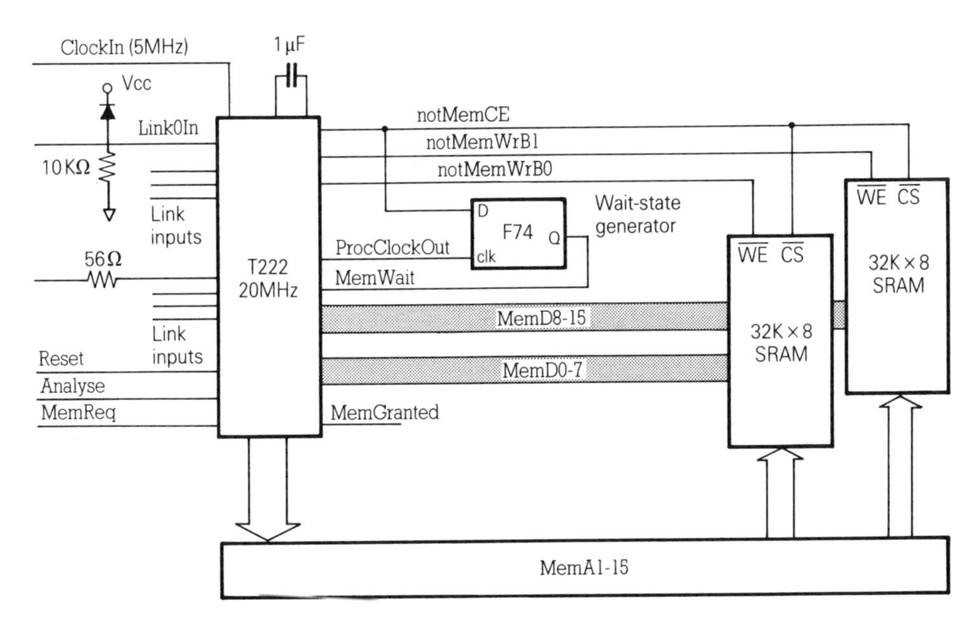

Figure 9.2 Sixteen-bit memory application

Table 9.2 T222-20 memory timing parameters

Symbol	Parameter	Min.	Max.
TAVEL	address valid before chip enable low	8	
TELEH	chip enable low period	68	80
TELEL	delay before chip enable reassertion	19	
TEHAX	address hold after chip enable high		
TELDrV	data valid from chip enable low	0	50
TAVDrV	data valid from address valid	0	63
TDrVEH	data setup before chip enable high	22	
TEHDrZ	data hold after chip enable high	0	20
TWEHEL	write enable setup before chip enable low	18	
TPCHEL	ProcClockOut high to chip enable low	8	
TDwVEH	data setup before chip enable high	50	
TEHDwZ	data hold after write	5	25
TDwZEL	write data invalid to next chip enable	1	
TWELEL	write enable setup before chip enable low	−8	
TEHWEH	write enable hold after chip enable high	−3	6

All parameters are in nsec

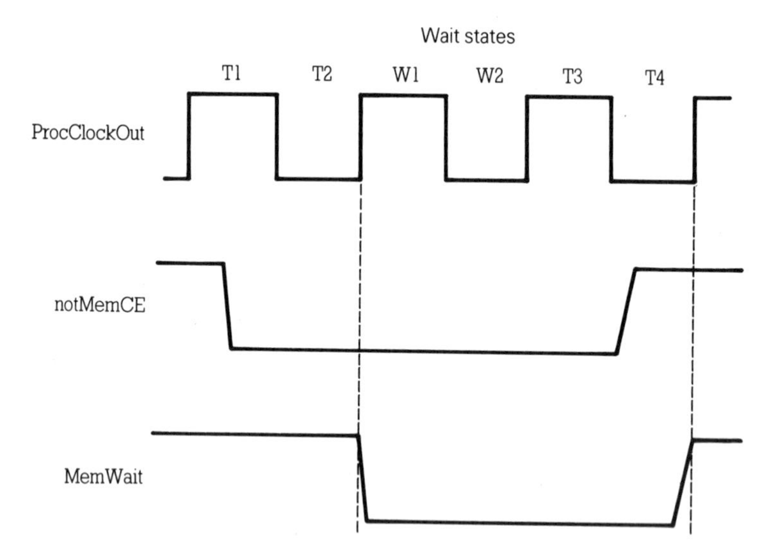

Figure 9.3 Wait-state generator timing

devices with faster access times can be used without wait states if external memory speed is important.

Direct access to the external memory by peripheral devices is provided for by the `MemReq` and `MemGranted` signals. The processor samples `MemReq` during every cycle of the processor clock when it is carrying out internal memory access. During external memory access `MemReq` is sampled in the first high phase of `ProcClockOut` after `notMemCE` goes low. `MemoryGranted` is asserted by the processor very quickly if an internal memory operation is in progress, with a delay of 85 to 100 nsec for a

20 MHz T212. Otherwise, the memory grant will be delayed until the end of the external memory cycle. Just how long this delay is depends on whether MemWait is being used to slow down external memory cycles. Before MemGranted goes active, the data and address bus, byte-write strobes and notMemCE will be put in a high impedance state. They will remain in that state until after MemReq is released and the processor has deasserted MemGranted. There is no upper limit to the length of time that MemReq may be held high. Thus DMA may be used by an external device to load a bootstrap program into memory immediately after reset, if MemReq is held high during the reset period. As soon as MemReq is released, the processor will access the first instruction of the bootstrap at location 0x7FFE.

9.2 Thirty-two bit programmable memory interface

One of the most interesting aspects of design using the 32-bit transputer is the programmable external memory interface. This interface can be used to add dynamic or static memory to a transputer system with almost no external logic. The interface can generate all the strobes required for dynamic memory (see Table 9.3), and contains a 10-bit refresh counter.

Table 9.3 Programmable memory interface signals

Signal	I/O	Function
MemAD2-31	in/out	multiplexed address and data bus
MemnotWrD0	in/out	data bit 0, multiplexed with write cycle flag
MemnotRfD1	in/out	data bit 1, multiplexed with refresh cycle flag
notMemS0	out	fixed memory strobe
notMemS1-4	out	programmable memory strobes
notMemRd	out	read strobe
notMemWrB0-3	out	byte-write strobes, programmable for early or late write
notMemRf	out	refresh indicator
RefreshPending	out	refresh is pending (T805 and T425 only)
MemWait	in	wait request
MemRequest	in	DMA request
MemGranted	out	DMA grant

The basic unit of memory timing is the unit Tm, which is half a cycle of the processor clock. This will be 25 nsec for a 20 MHz processor, 16.7 nsec for a 30 MHz device. The memory cycle is divided into six Tstates, T1 to T6, (see Table 9.4) each of which can be programmed to be 1–4 Tm long, giving a minimum external memory cycle time of three processor clock periods. The length of T4 can be extended indefinitely by the MemWait input.

Table 9.4 Tstates

T1	address setup before address valid strobe
T2	address hold time after address valid strobe
T3	read cycle tristate or write cycle data setup
T4	extendible data setup time
T5	read or write data
T6	data hold

9.2.1 Read and write cycles

Each Tstate corresponds to a different part of the memory access cycle (Table 9.3 and Figure 9.4). During T1 the memory address is placed on the multiplexed address and data bus, where it is held until the end of T2. The strobes notMemS0 and notMemS1 both go low at the end of T1, and so either can be used to latch the address. Strobe notMemS0 goes high again at the end of T5, the duration of notMemS1 is programmable. During T1 and T2 the lowest two bits of the address/data (AD) bus, MemnotWrD0 and MemnotRfD1 provide an early indication of a write or a refresh cycle.

In a read cycle both MemnotWrD0 and MemnotRfD1 will remain high until the transputer tristates the AD bus during T3. The read strobe notMemRd goes low at the end of T3. Data are read by the processor at the end of T5 when the read strobe goes high again. The transputer always reads a complete word of data; byte-read

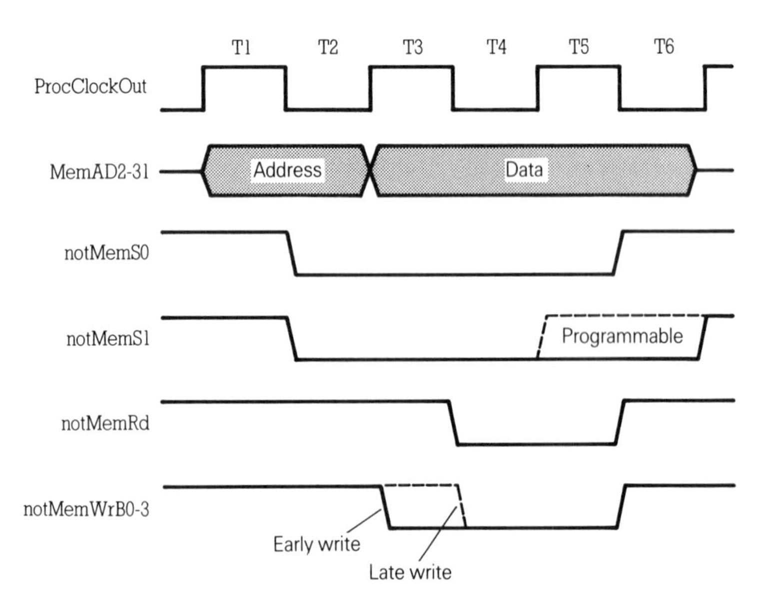

Figure 9.4 Thirty-two-bit memory access

accesses are managed internally. The duration of T4 can be extended indefinitely by holding MemWait high. This signal is sampled on a falling edge of ProcClockOut during T3, and if high T4 will be extended until MemWait goes low again.

In a write cycle MemnotWrD0 will be low during T1 and T2. The data are placed on the AD bus during T3, and held until the end of T6. Four byte-write strobes notMemWrB0-3 can be programmed to go low either during T3 (early write) or during T4 (late write); in either case the strobes go high again at the end of T5. These strobes should be used to enable writing into the memory devices of each byte, and the written data should be latched by the strobes going high. Again, the duration of T4, the data hold part of the write cycle, can be extended indefinitely by the MemWait input. This signal is sampled on a falling edge of ProcClockOut during T3, and if high T4 will be extended until MemWait goes low again.

A refresh cycle is indicated by the processor holding MemnotRfD1 low T1 and T2, and MemRF going low after MemnotRfD1. A refresh cycle (Figure 9.5) has the same timing as a normal read or write cycle except that two Tm periods are added before T1. The strobes notMemS0-4 are generated as normal, but both notMemRd and the byte-write strobes remain high. A 10-bit refresh address is put out on MemAD2-11; MemAD12-30 stay high but MemAD31 remains low.

The T805 and T425 have a RefreshPending output that indicates that the processor is about to perform a refresh cycle. The signal will remain active until the refresh cycle is started. This signal can be used by a DMA device to determine when refresh is due, and to suspend DMA while the processor carries out the refresh cycle.

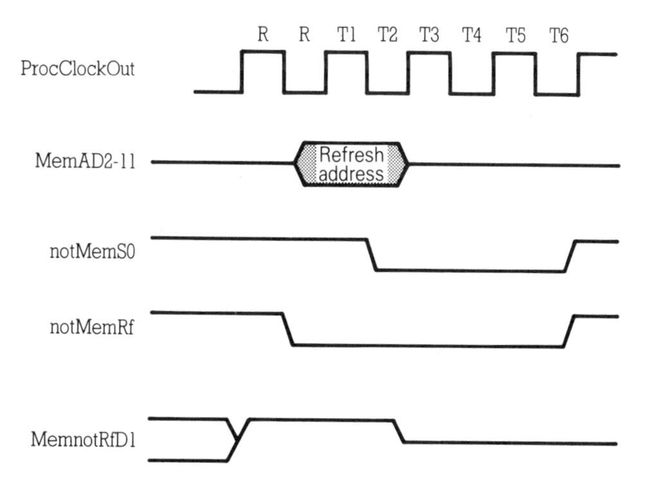

Figure 9.5 Refresh-cycle timing

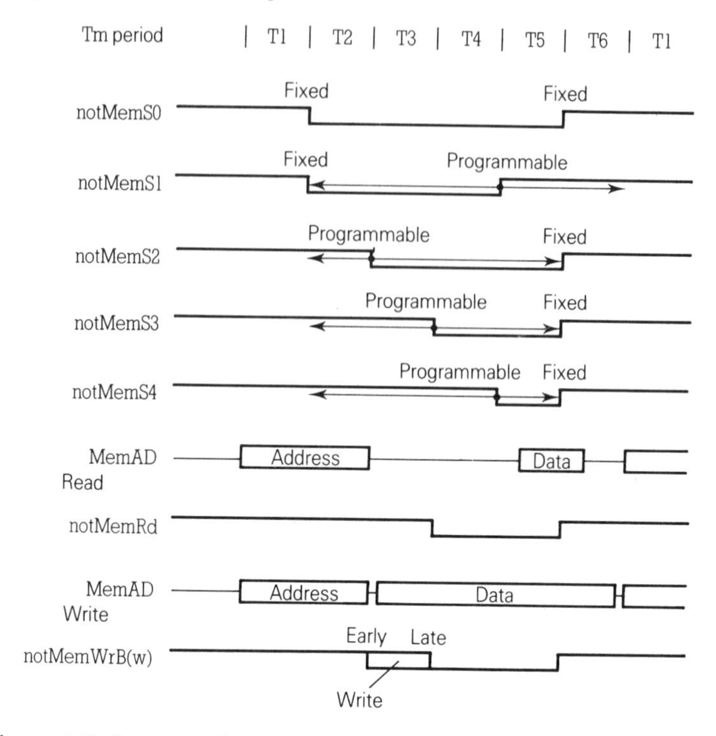

Figure 9.6 Summary of programmable strobes

9.2.2 Programmable strobes

The signals described above are usually sufficient for static memory applications, but dynamic memory will need to use the additional programmable strobes. These are summarized in Figure 9.6. The strobe `notMemS1` has been mentioned already. This always falls at the beginning of `T2`, but its duration is programmable from zero to thirty-one periods of `Tm`. However, it will rise at the end of `T6` even if configured to be longer. The three strobes `notMemS1-4` all go high at the end of `T5`, but can be programmed to go low from one to thirty-one `Tm` after the end of `T2`. A strobe programmed with a zero delay will never go low.

The data that configure the memory interface consist of 36 bits that are read in from the least significant bit of memory location `0x7FFFFF6C` to `0x7FFFFFF8` at reset time by the `MemConfig` pin. However, seventeen predefined configurations are supplied, and can be selected by connecting the `MemConfig` pin to one of the other pins of the memory interface. Table 9.5 shows the encoding of the memory configuration, and Table 9.6 the predefined configurations. It is very useful to have software that can display the memory cycle given the configuration data, and this is provided as part of the INMOS Transputer Development System, and also as a C program in Appendix B.

Table 9.5 Memory configuration encoding

0x7FFFFF6C	T1 lsb	0x7FFFFFB4	notMemS2 bit 1
70	T1 msb	B8	notMemS2 bit 2
74	T2 lsb	BC	notMemS2 bit 3
78	T2 msb	C0	notMemS2 bit 4
7C	T3 lsb	C4	notMemS3 bit 0
80	T3 msb	C8	notMemS3 bit 1
84	T4 lsb	CC	notMemS3 bit 2
88	T4 msb	D0	notMemS3 bit 3
8C	T5 lsb	D4	notMemS3 bit 4
90	T5 msb	D8	notMemS4 bit 0
94	T6 lsb	DC	notMemS4 bit 1
98	T6 msb	E0	notMemS4 bit 2
9C	notMemS1 bit 0	E4	notMemS4 bit 3
A0	notMemS1 bit 1	E8	notMemS4 bit 4
A4	notMemS1 bit 2	EC	RI lsb
A8	notMemS1 bit 3	F0	RI msb
AC	notMemS1 bit 4	F4	Refresh enable
B0	notmemS2 bit 0	F8	Late write

9.2.3 DMA

DMA can be requested at any time by taking `MemReq` high. This signal is sampled during `T6` of both refresh and external memory cycles, and every low period of `ProcClockOut` for internal memory access. The address bus is floated one processor clock period after the `ProcClockOut` rising edge which follows the sample. DMA devices can monitor the `RefreshPending` signal on the T805 and T425, and may suspend DMA when a refresh cycle is due. Otherwise, memory refresh is the responsibility of the DMA device.

9.3 Design example – a size 2 TRAM

In this section we will give the complete design of a size 2 T800 or T425 TRAM TRAM, with various options for processor and memory speed. The simplicity and low parts count of this design well illustrate the ease of hardware design with the transputer. Our design goals are as follows:

- T800 or T425 transputer, 20–30 MHz.
- 1–8 Mbytes dynamic RAM, 1 or 4 Mbit devices, slow or fast.
- Size 2 TRAM.
- Minimum parts count, maximum speed with devices used.

Table 9.6 Predefined memory configurations

Pin	T1	T2	T3	T4	T5	T6	S1	S2	S3	S4	Wr	RI	Tm
MemnotWrD0	1	1	1	1	1	1	30	1	3	5	1	72	3
MemnotRfD1	1	2	1	1	1	2	30	1	2	7	1	72	4
MemAD2	1	2	1	1	2	3	30	1	2	7	1	72	5
MemAD3	2	3	1	1	2	3	30	1	3	8	1	72	6
MemAD4	1	1	1	1	1	1	3	1	2	3	e	72	3
MemAD5	1	1	2	1	2	1	5	1	2	3	e	72	4
MemAD6	2	1	2	1	3	1	6	1	2	3	e	72	5
MemAD7	2	2	2	1	3	2	7	1	3	4	e	72	6
MemAD8	1	1	1	1	1	1	30	1	2	3	e	-	3
MemAD9	1	1	2	1	2	1	30	2	5	9	e	-	4
MemAD10	2	2	2	2	4	2	30	2	3	8	1	72	7
MemAD11	3	3	3	3	3	3	30	2	4	13	1	72	9
MemAD12	1	1	2	1	2	1	4	1	2	3	e	72	4
MemAD13	2	1	2	1	2	2	5	1	2	3	e	72	5
MemAD14	2	2	2	1	3	2	6	1	3	4	e	72	6
MemAD15	2	1	2	3	3	3	8	1	2	3	e	72	7
MemAD31	4	4	4	4	4	4	31	30	30	18	1	72	7

Descriptions:

MemnotWrD0	dynamic RAM in 3 processor cycles
MemnotRfD1	dynamic RAM in 4 cycles
MemAD2	dynamic RAM in 5 cycles
MemAD3	dynamic RAM in 6 cycles
MemAD4	multiplexed address dynamic RAM in 3 cycles
MemAD5	multiplexed address dynamic RAM in 4 cycles
MemAD6	multiplexed address dynamic RAM in 5 cycles
MemAD7	multiplexed address dynamic RAM in 7 cycles
MemAD8	fast static RAM in 3 cycles
MemAD9	static RAM in 4 cycles with wait generator
MemAD10	general purpose, 7 cycles
MemAD11	general purpose, 9 cycles
MemAD12	dynamic RAM in 4 cycles
MemAD13	dynamic RAM in 5 cycles
MemAD14	dynamic RAM in 6 cycles
MemAD15	dynamic RAM in 7 cycles
MemAD31	general purpose, 12 cycles

9.3.1 Initial design

We will assume that in order to achieve a minimum parts count we must use one of the predefined memory interface configurations. Which configuration we use will depend on the processor and memory speed, although we probably need not consider memory slower than 120 nsec or faster than 60 nsec access time. The TRAM size and parts count requirements suggest that either 256K $\times$ 4-bit or 1M $\times$ 4-bit dynamic memory devices should be used.

Fast 74ACT841 10-bit transparent latches can be used to latch the processor address at the end of `T1`. These devices have an output enable time of about 8 nsec, and the latch enable can be driven from `notMemS0`. Ten-bit column and row addressing allows the use of rams to 1M $\times$ 4-bit. The other memory control signals and latch output enables will be driven from a fast PAL, with a delay time of roughly 7 nsec. These considerations lead to the block diagram shown in Figure 9.7.

9.3.2 System services and links

The schematic of the processor, system services and link connections are shown in Figure 9.8. The `Reset` and `Analyse` signals are taken unchanged to TRAM input pins. `ErrorIn` is grounded, and `Error` connected to the base of a transistor which provides the open collector `notError` output. The `Clock` input is also taken directly from a TRAM pin; `ProcSpeedSelect2` is held high and `ProcSpeedSelect0` and `ProcSpeedSelect1` can be pulled high or low to select processor speeds between 20 and 30 MHz. The on-chip phase-locked loop is decoupled by a 1 μF capacitor connected between `CapPlus` and `CapMinus`; these connections must be by very short PCB tracks for correct operation.

The four link inputs are protected against static discharge by diodes to `Vcc`, and pulled low in the disconnected state by 10 Kohm resistors. The link outputs have 56 ohm series matching resistors. Link speed selection is provided by connecting `LinkSpeedA` on the TRAM to `Link0Special` on the transputer, and `LinkSpeedB` to `Link123Special` and pulling `LinkSpecial` high. This meets the TRAM specification which assumes that all links should be at either 10 or 20 Mbps, but it also allows `Link0` to be set at a different speed to the other links.

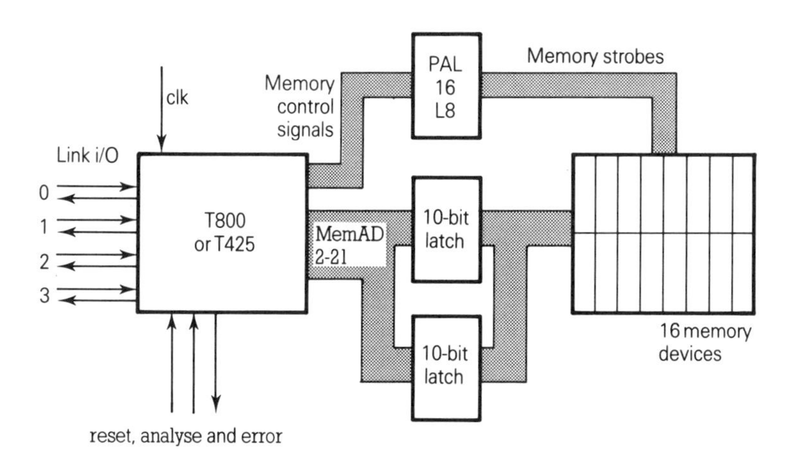

Figure 9.7 TRAM outline design

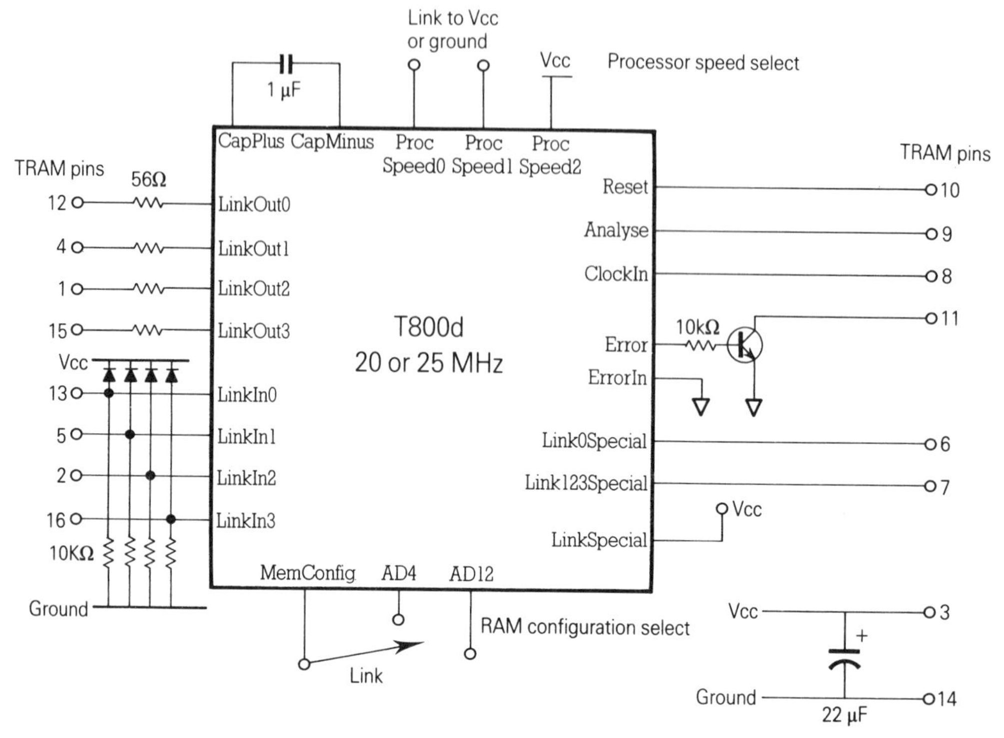

Figure 9.8 Processor, system services and link connections

9.3.3 Memory design

Figures 9.9 and 9.10 show the schematics of the basic memory design. Memory addresses are latched by `notMemS0` going low at the end of `T1`; all other memory control signals are generated by a fast PAL for maximum flexibility. Only one bank of memory is shown in Figure 9.10; the second bank is identical. Each bank has a separate $\overline{\texttt{ras}}$ signal. The bank is selected by address bit 11 for 256K × 4 dynamic ram devices and bit 22 for 1M × 4 devices. In refresh cycles both banks are active, but no $\overline{\texttt{cas}}$ occurs; refresh cycles are identified by `notMemRf` becoming active.

We now consider the design and timing of memory for a 20 MHz processor with four-cycle memory. Four cycles at 20 MHz is 200 nsec, and from Table 9.7 we can see that memory with an access time of at most 100 nsec can be used. There are two four-cycle configurations obtained by connecting `MemConfig` to either `AD5` or `AD12` (Figure 9.11). If we use `notMemS1` as the row address strobe, the `AD5` configuration gives a $\overline{\texttt{ras}}$ low time of 125 nsec, which is more than adequate, but a $\overline{\texttt{ras}}$ precharge time of only 75 nsec, which does not meet the memory specification. Therefore we must use the `AD12` configuration which gives approximately 100 nsec access time and 100 nsec precharge. This strobe configuration is not available on the T414.

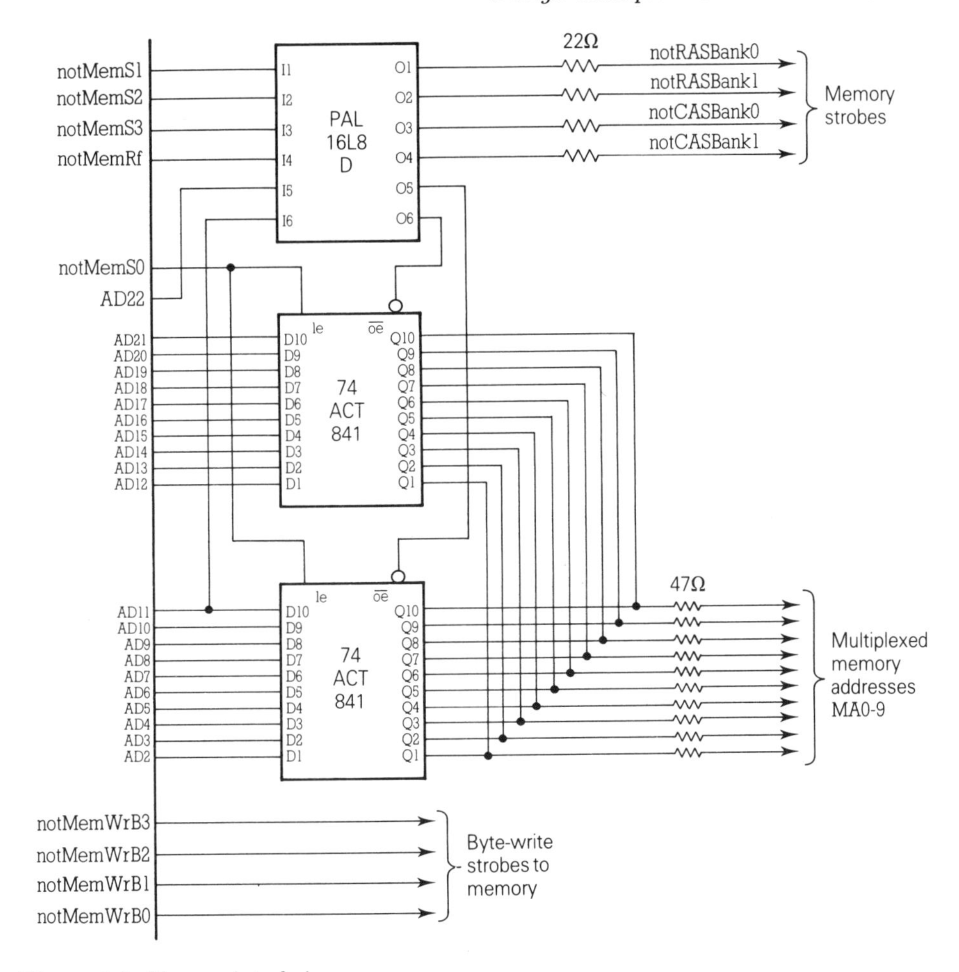

Figure 9.9 Memory interfacing

Table 9.7 Typical memory and transputer parameters

Access (nsec)	**ras** precharge (nsec)	Cycle time (nsec)	Number of processor cycles in memory cycle timing 20 MHZ	25 MHz	30 MHz
60	55	125	2.5	3.1	3.8
70	60	140	2.8	3.5	4.8
80	70	160	3.2	4.0	4.8
100	80	190	3.8	4.8	5.7
120	90	220	4.4	5.5	6.6

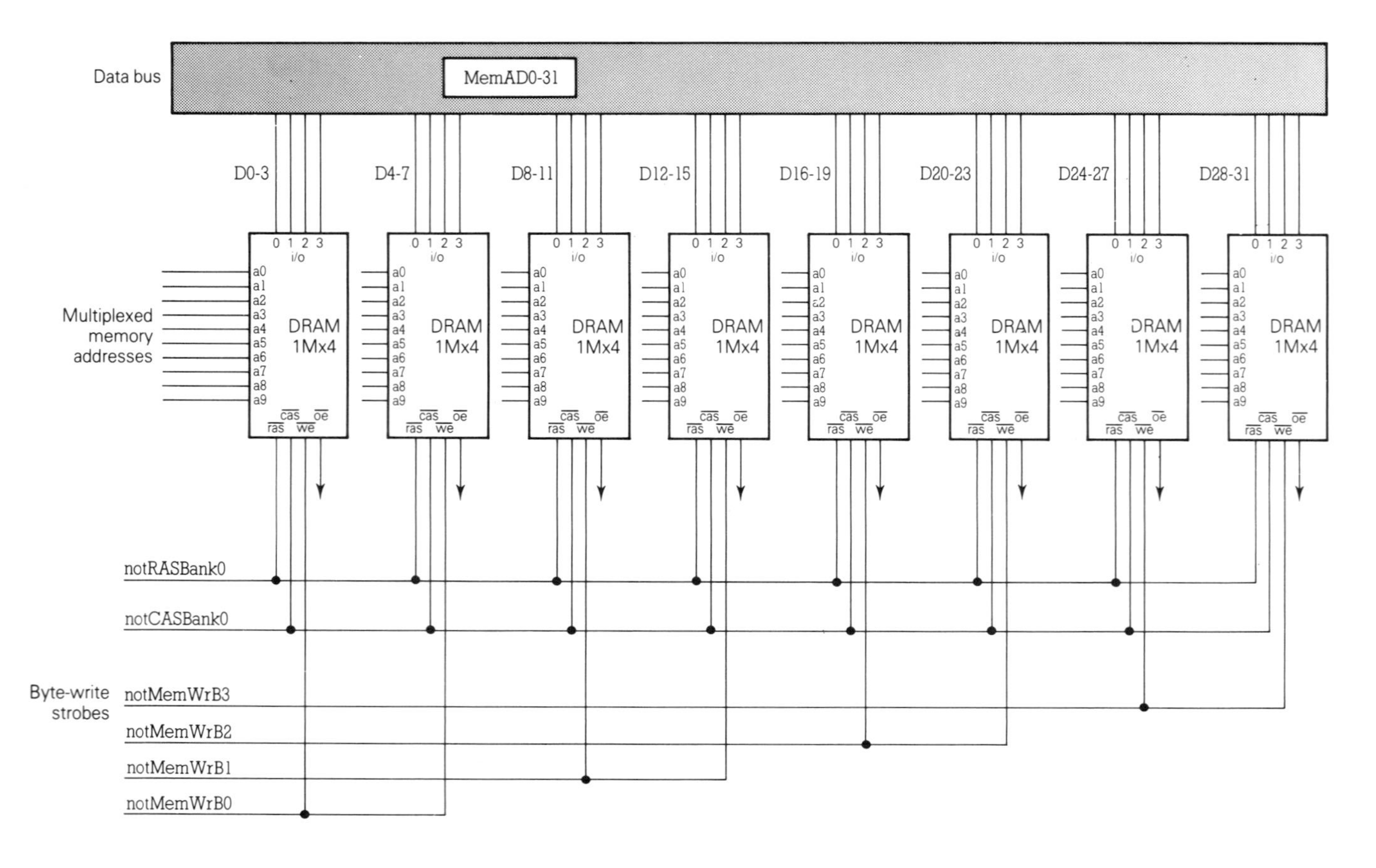

Figure 9.10 Memory, bank 0

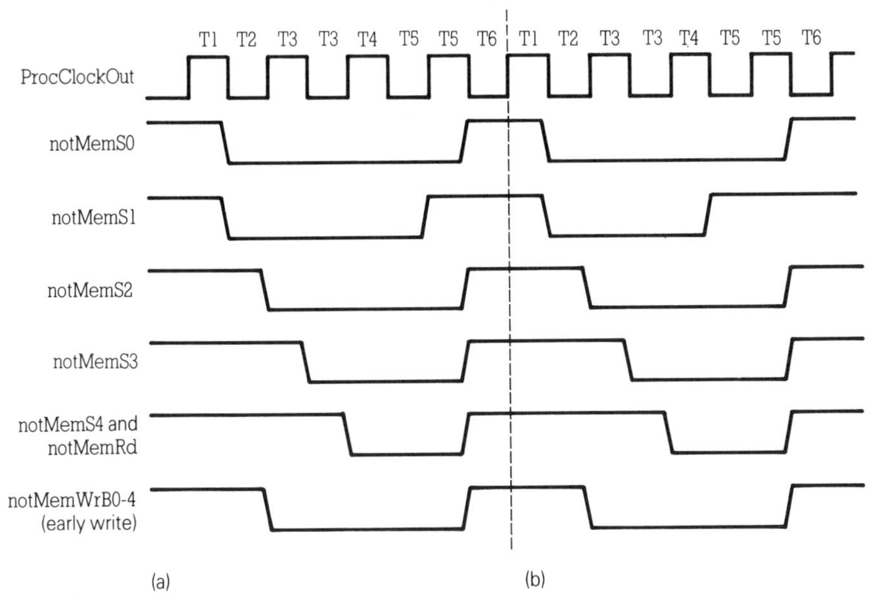

Figure 9.11 (a) AD5 and (b) AD12 memory strobe configurations

The strobe `notMemS2` can control column and row address multiplexing, and $\overline{\texttt{cas}}$ can be derived from `notMemS3`.

A complete timing diagram must take into account delays in the PAL and address latches, and the variations in strobe timing of the transputer. In the transputer specification the programmable strobe timing is related to `notMemS0`, and its timing is in turn related to `ProcClockOut`. Assuming accurate strobe timing and a 20 MHz processor leads to the timing diagram of Figure 9.12.

The timing requirements for full-speed memory with a 25 MHz processor are much tighter. Three processor cycles at 25 MHz processor takes 120 nsec, and Table 9.7 suggests that it is just possible to achieve this memory timing with 60 nsec access time devices. The only three cycle configuration is `AD4`, which gives 60 nsec access and 60 nsec precharge times. Similarly, using `notMemS3` as `cas` gives a low time of 40 nsec and a hold time for $\overline{\texttt{ras}}$ after $\overline{\texttt{cas}}$ goes low of 20 nsec. This is just within the published specification of the Hitachi HM41256H devices. The complete timing diagram is shown in Figure 9.13; clearly the tolerances are a little lower than would be hoped for in a truly reliable design, but TRAMs have been constructed to this specification and work well.

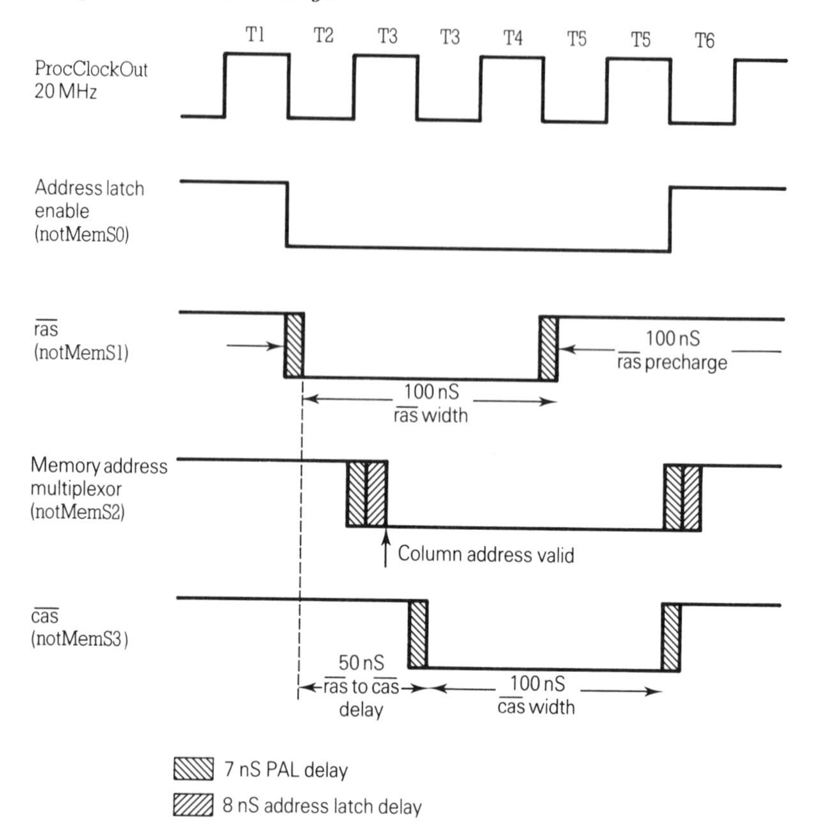

Figure 9.12 Memory timing for AD12 configuration with 20 MHz processor

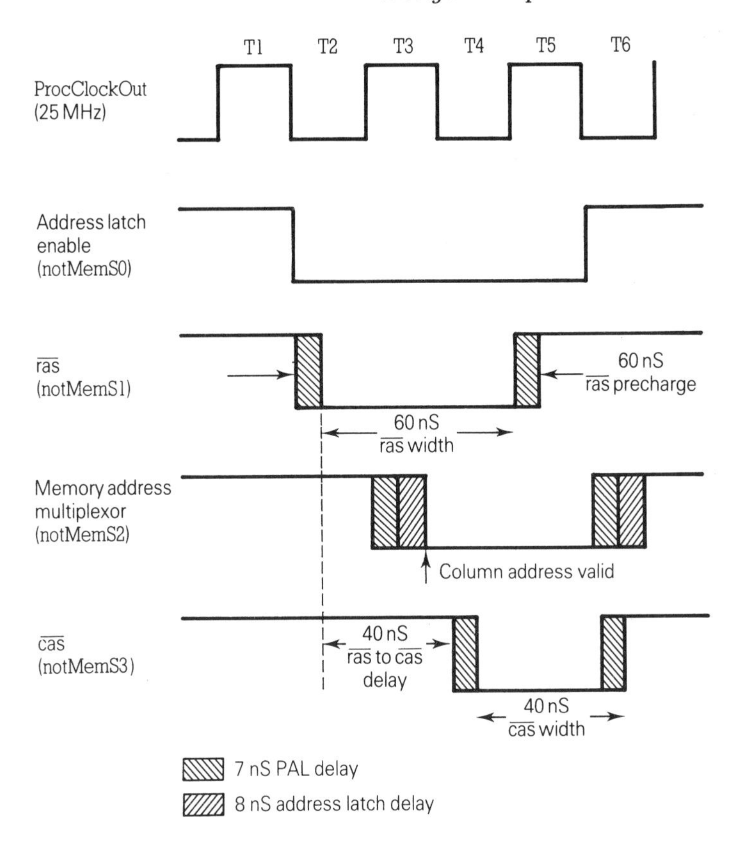

Figure 9.13 Memory timing for AD4 configuration with 25 MHz processor

Appendix A

Instruction Set Reference

In this appendix we present an informal specification of the transputer instruction set. The instructions have been tabulated in functional groups, and some instructions will appear in more than one table. For each instruction we give its 'opcode'. For direct functions this will be the function nibble of the instruction byte, for other integer processor instructions it is the value that must be loaded into the operand register O when the `opr` function is executed. Some floating-point instructions are executed by the **`fpuentry`** instruction, which interprets an 'opcode' that has been loaded in the A register.

The next column of each table contains the mnemonic of the instruction, and the third column timing information. This timing information assumes that any non-register operands are in internal (one-cycle) memory; the timings for external memory operands vary in a rather complex way which is described in the performance section of the engineering data of each processor. Differences in timings for 16- and 32-bit processors are indicated in the few cases where this is necessary.

The fourth column of each table indicates if an instruction is a potential descheduling point (D), or can affect the value of the processor error flag (E) or the floating-point unit error flag (F). The fifth column is an informal description of the action of the instruction.

Constants

Constant	16-bit value	32-bit value	Description
`MinInt`	0x8000	0x80000000	minimum integer value
`BytesPerWord`	2	4	number of bytes in a word
`TRUE`	1	1	representation of logical true
`FALSE`	0	0	representation of logical false

Prefix, operate and direct functions

Prefix

`0x2`	`pfix`	1	prefix; load operand into lowest 4 bits of O then shift O up 4 places
`0x6`	`nfix`	1	negative prefix; load operand into lowest 4 bits of O, complement O then shift O up 4 places

Operate

`0xF`	`opr`	operate; load operand into lowest 4 bits of O, then start the execution of the instruction whose opcode is in O

Direct

`0x0`	`j`	3	D	jump; set $I = I + O$, A, B and C become undefined (see also debugging instructions)
`0x1`	`ldlp`	1		load pointer to local; load A with the address of the word offset by O from W
`0x3`	`ldnl`	2		load non-local variable; load A with the word offset by O from A
`0x4`	`ldc`	1		load constant; sets $A = O$, a large constant can be loaded by a series of `pfix` or `nfix` operations preceding the `ldc`
`0x5`	`ldnlp`	1		load pointer to non-local; load A with the address of the word offset by O from A
`0x7`	`ldl`	2		load local; load A with the word offset by O from W
`0x8`	`adc`	1	E	add constant; set $A = A + O$
`0x9`	`call`	7		call; set $W = W - 4$
`0xa`	`cj`	2		(not taken) conditional jump;
		4		(taken); if $A = 0$ then set $I = I + O$ else pop A
`0xB`	`ajw`	1		adjust workspace; set $W = W + O$
`0xC`	`eqc`	2		equals constant; if $A = O$ set $A = 1$ else set $A = 0$
`0xD`	`stl`	1		store local; store the value of A at the word offset by O from W
`0xE`	`stnl`	2		store non-local; store the value of B in the word offset by O from A

Notes: The direct functions load the data part of the instruction byte into the lowest 4 bits of the O register before execution. All functions except `pfix` and `nfix` clear the operand register after execution.

In the load and store instructions the O register must contain a word offset (rather than a byte offset), and the base register, A or W, must point to a word boundary.

Arithmetic/logical instructions

0x46	and	1		bitwise AND; $A = A \& B$
0x4B	or	1		bitwise OR; $A = A \mid B$
0x33	xor	1		bitwise exclusive OR; $A = A \oplus B$
0x32	not	1		bitwise NOT; $A = \sim A$
0x41	shl	n+2		shift left B by A places
0x40	shr	n+2		shift right B by A place
0x05	add	1	E	add; $A = A + B$, checked
0x0C	sub	1	E	subtract; $A = B - A$, checked
0x53	mul	38	E	multiply, 32-bit processor
		23	E	multiply, 16-bit processor; $A = A * B$, checked
0x72	fmul	35	E	fractional multiply; no rounding
		40	E	fractional multiply, rounding
0x2C	div	39	E	divide; 32-bit processor
		24	E	divide, 16-bit processor; set $A = B/A$, error flag is set if A is zero or if $A = -1$ and $B =$ MinInt since the result would be -MinInt which cannot be represented
0x1F	rem	37	E	remainder, 32-bit processor
		21		remainder, 16-bit processor; generates the remainder after dividing B by A, sets error flag under the same conditions as div
0x09	gt	2		greater than; if $B > A$ then set $A = 1$ else set $A = 0$
0x04	diff	1		unsigned subtraction; set $A = B_{unsigned} - A_{unsigned}$
0x52	sum	1		unchecked addition; set $A = B_{unsigned} + A_{unsigned}$
0x08	prod	b+4		product; $A = A * B$, unchecked, b is the position of the most significant bit (msb) of A
		m+5		product for negative register A. The fast implementation of prod for negative A is not used on the T414, T222 or T212

Notes: In all the above instructions except not the values of the operands are taken from A and B. The result is stored in A, the value of C is popped into B and the new value of C is undefined.

Long arithmetic instructions

0x16	ladd	2	E	long add; $A = (B + A) + C_{lsb}$, set error flag if carry is generated
0x38	lsub	2	E	long subtract; $A = (B - A) - C_{lsb}$, set error flag if borrow is generated
0x37	lsum	3		long sum; form $(B + A) + C_{lsb}$. Load least significant word of result into A and most significant (carry) bit into B
0x4F	ldiff	3		long difference; form $(B - A) - C_{lsb}$. Load least significant word of result into A and borrow bit into B
0x31	lmul	33		long multiply; form $(A * B) + C_{lsb}$. Load least significant word of result into A, most significant into B
0x1A	ldiv	35	E	long divide; divide double length unsigned value in B and C (most significant word (msw) in C) by single length unsigned value in A. Load A with result, B with remainder. Overflow causes error flag to be set
0x36	lshl	n+3		long shift left, shift double word in C and B, A places to the left. A contains least significant word (lsw) and B msw of the result
		n-28		long shift left
0x35	lshr	n+3		long shift right, as `lshl`, except shift right A places
		n-28		long shift right
0x19	norm	n+5		normalize
		n-26		normalize $n > 32$
		3		normalize, $n = 64$

Notes: In the long arithmetic instructions the operands are taken from the three evaluation stack registers, the lsw of the result is placed in A and the msw in B.

General instructions

0x00	rev	1		exchange values of A and B
0x3A	xword	4		sign extend the part word in B to a whole word. The part word is indicated by setting its most significant bit in A
0x56	cword	5	E	check word; if $B \geq A \vee B < -A$ then set *ErrorFlag*, checks that the value in B can be represented in a bit field whose msb is set in A
0x1d	xdble	2		extend to double
0x4C	csngl	3	E	check single; converts from a double length integer in A and B, msw in B, to a single length integer in A. *ErrorFlag* is set if there are significant bits in B
0x42	mint	1		load A with `MinInt`
0x5A	dup	1		duplicate top of stack; sets $A = A$, $B = A$ and $C = B$; T805, T801, T425 and T225 only
0x79	pop	1		rotates evaluation stack; sets $A = B$, $B = C$ and $C = A$; T805, T801, T425 and T225 only

Two-dimensional block move instructions

0x5B	move2dinit	8	the source stride in C, the destination stride in B and number of rows to copy in A are set up for one of the `move2d` instructions
0x5C	move2dall	*	performs a two-dimensional block copy from the array given in C to the array given in B, A contains the number of bytes in each row to copy, other parameters must have already been set up by `move2dinit`
0x5D	move2dnonzero	*	two-dimensional block copy of non-zero bytes
0x5E	move2dzero	*	two-dimensional block copy of zero bytes

Notes: See Section 3.13. Only in T805, T801, T800 and T425. Timing is very complex and best discovered by measurement in the actual system.

CRC and bit manipulation instructions

0x74	`crcword`	35	calculate CRC on word
0x75	`crcbyte`	11	calculate CRC on byte
0x76	`bitcnt`	b+2	count bits set in word; $A = B + count$, the result in A is the sum of the number of bits set in A plus the value in B, so that the number of bits set in a sequence of words may be totalled. The time taken by the instruction depends on the position of the msb of A
0x77	`bitrevword`	36	reverse the bit pattern in A
0x78	`bitrevnbits`	n+4	the bottom A bits of B are reversed and the result left in A with the more significant bits zeroed

Notes: Only on T805, T801, T800, T425 and T225 processors. In the CRC instructions the A register holds the word or byte to be combined with the accumulated CRC in B, using the generator in C. For `crcbyte` the byte to be combined must be the top byte of A.

Indexing/array instructions

0x02	bsub	1	byte subscript; load A with the address of the byte offset by B from the base address in A. This acts as an unchecked addition of A and B in present transputers
0x0A	wsub	2	word subscript; load A with the address of the word offset by B words from the base address in A
0x81	wsubdb	3	double word subscript; load A with the address of the double word offset by B double words from the base address in A. T805, T801, T800 and T425 only
0x34	bcnt	2	byte count; sets $A = A *$ `BytesPerWord`, converting a length in words into a length in bytes. Helpful in writing code that is independent of processor word length
0x3F	wcnt	5	word count; decomposes an address in A into a word address, returned in A, and a byte selector, returned in B
0x01	lb	5	load byte; loads a single byte from the address contained in A, the most significant bytes of A are zeroed
0x3B	sb	4	store byte; stores the least significant 8 bits of B in the byte pointed to by A
0x4A	move	2w+8	move message; copy the number of bytes given in A from the starting at C to the block starting at B, if the blocks overlap the result is undefined

Timer instructions

0x22	ldtimer	2		loads the current value of the timer into A
0x2b	tin	30	D	(time future)
		4		(time past)
				waits until the current priority time is AFTER the time value in A. If the required time has already passed the process continues, but if not the process is placed on the timer queue
0x4E	talt	4		timer ALT start – see ALT instructions
0x51	taltwt	15	D	timer ALT wait (time past) – see ALT instructions
		48	D	timer ALT wait (time future)
0x47	enbt	8		enable timer – see ALT instructions
0x2E	dist	23		disable timer – see ALT instructions

Notes: Which timer is used in these instructions depends on the priority of the process executing the instruction. The high-priority timer ticks every microsecond, the low-priority timer every 64 μsec.

Input/output instructions

0x07	in	2w+19	D	input message
0x0B	out	2w+19	D	output message
0x0F	outword	23	D	output word
0x0E	outbyte	23	D	output byte
0x12	resetch	3		reset channel

Notes: See Section 3.8.

ALT instructions

0x43	alt	2		ALT start
0x44	altwt	5	D	ALT wait (channel ready)
		17	D	ALT wait (channel not ready)
0x45	altend	4		ALT end
0x49	enbs	3		enable skip
0x30	diss	4		disable skip
0x48	enbc	7		enable channel (ready)
		7		enable channel (not ready)
0x2F	disc	8		disable channel

Notes: See Section 3.10.

Flow of control instructions

0x9	`call`	7		call; call a procedure whose entry point is offset O bytes the next instruction; set $W = W - 4$, $W[0] = I$, $W[1] = A$, $W[2] = B$, $W[3] = C$, $A = I$ and $I = I + O$. Direct function
0x20	`ret`	5		return; return from a procedure call by loading I from $W[0]$ and incrementing W by 4 words. This reverses the action of `call`; the values of A, B and C are not affected, allowing up to three results to be returned
0x1B	`ldpi`	2		load pointer to instruction; set $A = I + A$, loads the address of instruction offset by A from I
0x3C	`gajw`	2		general adjust workspace; swaps values of A and W, allows W be set to an arbitrary value
0x06	`gcall`	4		general call; swaps values of A and I, allows the use of any form of calling sequence.
0x21	`lend`	10	D	loop end (loop)
		5	D	loop end (done)
				A loop is controlled by two contiguous words of memory pointed to by B, the first word containing the loop variable, the second the number of iterations left to perform. This instruction decrements the iteration count, and if the result is greater than zero increments the loop variable and branches to the instruction whose offset is given in A which is positive for backward jumps and negative for forward jumps; set $B[1] = B[1] - 1$; if $B[1] > 0$ then set $B[0] = B[0] - 1$ and $I = I - A$

Scheduling instructions

0x0D	`startp`	12	D	start process
0x03	`endp`	13	D	end process
0x39	`runp`	10		run process
0x15	`stopp`	11		stop process
0x1E	`ldpri`	1		load current priority

Notes: See Section 3.7.

Error handling instructions

`0x13`	`csub0`	2	E	check subscript from 0; if $B_{unsigned} \geq A_{unsigned}$ then set error flag. Pops stack once
`0x4D`	`ccnt1`	3	E	check count from 1
`0x29`	`testerr`	2		test error false and clear (no error)
		3		test error false and clear (error)
`0x10`	`seterr`	1	E	set error
`0x55`	`stoperr`	2	D	stop on error (no error)
`0x57`	`clrhalterr`	1		clear *HaltOnErrorFlag*
`0x58`	`sethalterr`	1		set *HaltOnErrorFlag*
`0x59`	`testhalterr`	2		test *HaltOnErrorFlag*

Notes: See Section 3.11.

Processor initialization instructions

`0x2A`	`testpranal`	2	test processor analyzing
`0x3E`	`saveh`	4	save high-priority queue registers
`0x3D`	`savel`	4	save low-priority queue registers
`0x18`	`sthf`	1	store high-priority front pointer
`0x50`	`sthb`	1	store high-priority back pointer
`0x1C`	`stlf`	1	store low-priority front pointer
`0x54`	`stlb`	1	store low-priority back pointer
`0x17C`	`lddevid`	1	load A with the device identity: T425 0–9; T805 10–19; T801 20–29; T225 40–49. This instruction is a `NOP` on the T414, T222 and T212 processors, on the T800 it sets $B = C$ and leaves A undefined.
`0x7E`	`ldmemstartval`	1	load A with the value of `MemStart`. Only on T805, T425 and T225 processors

Notes: These instructions will normally only be used in the bootstrapping sequence.

Debugger support instructions

`0x0`	`jump 0`	3	D	jump 0 (break not enabled)
		11		jump 0 (break enabled, high-priority)
		13		jump 0 (break enabled, low-priority)
`0xB1`	`break`	9		break (high-priority)
		11		break (low-priority)
`0xB2`	`clrj0break`	1		clear *EnableJ0Break* flag
`0xB3`	`setj0break`	1		set *EnableJ0Break*
`0xB4`	`testj0break`	2		test *EnableJ0Break* set
`0x7A`	`timerdisableh`	1		disable high-priority timer interrupt
`0x7B`	`timerdisablel`	1		disable low-priority timer interrupt
`0x7C`	`timerenableh`	6		enable high-priority timer interrupt
`0x7B`	`timerenablel`	6		enable low-priority timer interrupt

Notes: See Section 3.15. On T805, T801, T425 and T225 only.

Floating-point support instructions

0x73	cflerr	3	E	sets *ErrorFlag* if A contains floating-point infinity or Not a Number (NaN)
0x63	unpacksn	15		unpacks single length floating-point number in A, returning exponent in B, fraction in A. The fraction does not include the implied msb of a normalized floating-point number. The C register is loaded with $4 * B$ plus a flag value of 0 if $A = 0$, 1 if A contained a normalized or unnormalized floating-point number, 2 if A was a floating-point infinity, 3 if A is NaN
0x6D	roundsn	12/15		round single-length floating-point number
0x6C	postnormsn	5/30		post-normalize correction of single-length floating-point number
0x71	ldinf	1		load A with 32-bit floating-point infinity
0x9C	fptesterr	1		On processors with hardware floating-point this instruction serves to load the A register with 1 if the floating-point error flag is set. On the T425 and T400 the instruction will always load A with 1, making it possible to test for the presence of a hardware floating-point unit by resetting the error flag and then executing this instruction. This instruction was not implemented on the T414.

Notes: See Section 3.14. These instructions are available only on the T425 and T414 processors, i.e. those 32-bit processors without a floating-point unit. The purpose of these instructions is to facilitate the software implementation of IEEE-754 standard floating-point arithmetic. The `norm`, `postnormsn` and `roundsn` instructions are used in code sequences that convert integers to floating-point numbers.

Floating-point instructions

Entry code and miscellaneous instructions

`0xAB`	`fpentry`	1	floating-point entry; starts the execution of the floating-point instruction whose opcode is in A
`0xA4`	`fprev`	1	floating-point reverse; swap contents of FA and FB
`0xA3`	`fpdup`	1	floating-point duplicate; push FB into FC, set $FB = FA$

Notes: See Section 3.14.

Load/store instructions

`0x8E`	`fpldnlsn`	2		floating-point load non-local single
`0x8A`	`fpldnldb`			floating-point load non-local double
`0x86`	`fpldnlsni`	4		floating-point load non-local indexed single
`0x82`	`fplsnldbi`	6		floating-point load non-local indexed double
`0x9F`	`fpldszerosn`	2		load zero single
`0xA0`	`fpldzerodb`	2		load zero double
`0xAA`	`fpldnladdsn`	8/11	F	floating-point load non-local and add single
`0xA6`	`fpldnladddb`	9/12	F	floating-point load non-local and add double
`0xAC`	`fpldnlmulsn`	13/20	F	floating-point load non-local and multiply single
`0xA8`	`fpldnlmuldb`	21/30	F	floating-point load non-local and multiply double
`0x88`	`fpstnlsn`	2		floating-point store non-local single
`0x84`	`fpstnldb`	3		floating-point store non-local single
`0x9E`	`fpstnli32`	4		store non-local int32

Notes: The floating-point load and store instructions all use A as the source or target address. For an indexed load B contains the index value, which will be used as a word or double word offset, depending on the size of the object loaded. The 'load and ...' instructions are short versions of common instruction sequences. Each load instruction sets a flag associated with the FA register to indicate the length of the object that has been loaded. The arithmetic and store instructions check this flag, so that different instructions are not needed for 32- and 64-bit operations. The timings shown are in the form typical/maximum cycles.

Arithmetic instructions

0x87	fpadd	6/9	6/9	F	floating-point addition
0x89	fpsub	6/9	6/9	F	floating-point subtraction
0x8B	fpmul	11/18	18/27	F	floating-point multiplication
0x8C	fpdiv	16/28	31/43	F	floating-point division
0x0B*	fpuabs	2/2	2/2	F	replace FA by its absolute value
0x8F	fpremfirst	36/46	36/46	F	floating-point remainder first step
0x90	fpremstep	32/36	32/36		floating-point remainder iteration
0x01*	fpusqrtfirst	27/29	27/29	F	floating-point square root first step
0x02*	fpusqrtstep	42/42	42/42		floating-point square root iteration
0x03*	fpusqrtlast	8/9	8/9		floating-point square root last step
0x0A*	fpuexpinc32	6/9	6/9	F	$FA = FA * 2^{32}$
0x09*	fpuexpdec32	6/9	6/9	F	$FA = FA/2^{32}$
0x12*	fpumulby2	6/9	6/9	F	$FA = FA * 2.0$
0x11*	fpudivby2	6/9	6/9	F	$FA = FA/2.0$

Notes: Opcodes suffixed by * are the values that have to be loaded into the A register before the `fpentry` instruction is executed. Timings shown are typical/maximum processor cycles for 32-bit (column 3) and 64-bit (column 4) operands.

The four binary arithmetic operations calculate $FB op FA$, leaving the result in FA and popping FC into FB. The new value of FC is undefined.

The remainder and square root operations are carried out by code sequences, for details see Section 3.14. The remainder is calculated from a divisor in FA and dividend in FB, the result is returned in FA. The contents of FC will be lost. Each remainder instruction pushes a boolean into A to control looping, thus the contents of the integer evaluation stack will also be lost. In `fpsqrt` FA contains the input and result, but FB and FC are used in the calculations and their previous contents lost. The iteration step `fpsrtstep` must be carried out twice for a 32-bit number, five times for a 64-bit number.

Rounding mode instructions

0x22*	fpurn	1	set rounding mode to Round_to_Nearest
0x06*	fpurz	1	set rounding mode to Round_to_Zero
0x04*	fpurp	1	set rounding mode to Round_to_Plus_Infinity
0x05*	fpurm	1	set rounding mode to round Round_to_Minus_Infinity

Notes: The floating-point load instructions set the rounding mode to Round_to_Nearest, this is the IEEE-754 default mode. To use another mode it is necessary to set it explicitly before each arithmetic operation.

Comparison instructions

0x94	fpgt	4/6	F	floating-point greater than
0x95	fpeq	3/5	F	floating-point equality
0x92	fpordered	3/4		floating-point orderability
0x91	fpnan	2/3		floating-point NaN
0x93	fpnotfinite	2/2		floating-point not finite
0x0E*	fpuchki32	3/4	F	check in range of type int32
0x0F*	fpuchki64	3/4	F	check in range of type int64

Note: Timings shown are typical/maximum processor cycles.

Conversion instructions

0x07*	fpur32tor64	3/4	F	real32 to real64
0x08*	fpur64tor32	6/9	F	real64 to real32
0x9D	fprtoi32	7/9	F	real to int32
0x96	fpi32tor32	8/10		int32 to real 32
0x98	fpi32tor64	8/10		int32 to real64
0x9A	fpb32tor64	8/8		bit32 to real64
0x0D*	fpunoraound	2/2		real64 to real32, no round
0xA1	fpint	5/6	F	round to floating integer

Note: Timings shown are typical/maximum processor cycles.

Error-handling instructions

0x83	fpchkerror	1	E	load *ErrorFlag* with the OR of *ErrorFlag* and *FpErrorFlag*
0x9C	fptesterror	2	F	set *A* to TRUE if *FpErrorFlag* not set, and clear *FpErrorFlag*
0x23*	fpuseterror	1	F	set *FpErrorFlag*
0x9C	fpuclearerror	1	F	clear *FpErrorFlag*

Appendix B

Memory Configuration Program

The memory configuration program takes as its input a set of specifications for the duration of the Tstates and the timing of the strobes. These are in the order:

```
T1 T2 T3 T4 T5 T6
S1 S2 S3 S4
L W
```

where the `Tn` gives the duration of each Tstate, `S1` is the length of strobe `notMemS1`, and `S2-4` are the delays of strobes `notMemS2-4` after the start of `T2`. All of these parameters are in units of `Tm` (half-cycles of `ProcClockOut`). If any strobe coefficient is specified as zero, this strobe will not go low.

The next non-blank character indicates early or late byte-write strobes, an upper or lower case 'L' showing late write, any other character early write. The final input parameter is an integer specifying the number of wait states to be added to `T4`.

The name of the file containing the configuration is passed to the program in the parameter line, and the configuration details are read in by the function `read_config`. The Tstate durations are calculated in `set_tstates` where the wait states are added into `T4`. If wait states would cause `T5` to start on a falling edge of `ProcClockOut` an extra `Tm` is added to `T4`. The duration of the complete memory cycle must always be an even number of `Tm` and an extra `Tm` will be added to `T6` if necessary.

The function `set_strobes` computes the starting and ending point of each programmable strobe, and of the read and write strobes. The results are printed by the function `print_data`, which reports the starting and ending points of each Tstate and strobe in units of `Tm` from the beginning of the memory cycle. It is easy to use these results to drive a program to plot the memory cycle, but this is not included here as the code is so device dependent.

For example, the program input corresponding to the AD12 predefined configuration is as follows:

```
1 1 2 1 2 1
4 1 2 3
e 0
```

The output produced by the program is then:

```
Transputer memory configuration program
Configuration file: cfg/ad12.cfg

T state durations
    T1      T2      T3      T4      T5      T6
    1       1       2       1       2       1

Strobe coefficients
    S1      S2      S3      S4
     4       1       2       3
Early write
0 wait states inserted into T4

   Tstates
    start      end

T1   0          1
T2   1          2
T3   2          4
T4   4          5
T5   5          7
T6   7          8


 Strobes
          start       end

notMemS0    1          7
notMemS1    1          5
notMemS2    2          7
notMemS3    3          7
notMemS4    4          7
notMemRd    4          7
notMemWr    2          7
```

```
#include <stdio.h>
#include <ctype.h>
#define TRUE 1
#define FALSE 0

struct T { int end; int duration; };
struct S { int start, end, data; };

char *config_name;

/* global variables describing the configuration   */
struct T tstate[7];           /* Tstate information */
struct S strobe[5];           /* strobe coefficients */
int waits;                    /* number of wait states inserted */
int early_write;              /* boolean for early or late write */
struct S notMemRd, notMemWr; /* read and write strobes */

void read_config(char *config_name)
{
FILE *c_file;
int i, c = ' ';
c_file = fopen(config_name, "r");
printf("\n Configuration file: %s\n",config_name);

/* get tstate durations */
for (i = 1; i < 7; i++) fscanf(c_file,"%d",&tstate[i].duration);
printf("\nT state durations \n     T1     T2     T3     T4     \
T5     T6\n");
for (i = 1; i < 7; i++) printf(" %5d ",tstate[i].duration);

/* get strobe coefficients */
for (i = 1; i < 5; i++) fscanf(c_file,"%d",&strobe[i].data);
printf("\nStrobe coefficients\n    S1     S2     S3     S4\n");
for (i = 1; i < 5; i++) printf(" %5d ",strobe[i].data);

/* get early or late write flag */
while (isalpha(c) == 0)
     c = fgetc(c_file);    /* get first non-blank char*/
if (c == 'L' || c == 'l')
           { early_write = FALSE;
             printf("\n Early write \n");
           }
   else    { early_write = TRUE;
             printf("\n Late write \n");
           }

fscanf(c_file,"%d",&waits);
printf(" %d wait states inserted into T4\n", waits);
}
```

```
void set_tstates()
/* compute the end point of each tstate */
{ int i;

/* add the wait states into T4 */
tstate[4].duration = tstate[4].duration + waits;
tstate[0].end = 0;
for (i=1; i <5; i++)
   tstate[i].end = tstate[i-1].end + tstate[i].duration;

/* add in an extra wait state to T4 if T5 would have to begin on
   a falling clock edge */
if (waits != 0 && ( tstate[4].end % 2) == 1)
  { tstate[4].end = tstate[4].end + 1;
    tstate[4].duration = tstate[4].duration + 1;
    waits = waits + 1;
    printf("Extra wait state added\n");
  }

for (i = 5; i <7; i++)
    tstate[i].end = tstate[i-1].end + tstate[i].duration;

/* If the total number of T's is odd, one must be added to T6 */
if ((tstate[6].end % 2 ) == 1)
     { tstate[6].end = tstate[6].end + 1;
       tstate[6].duration = tstate[6].duration + 1;
       printf("Extra Tm added to T6\n");
     }
}


void set_strobes()
{ int i;

/* notMemS0, fixed format strobe, from end of T1 to end of T5 */
strobe[0].start = tstate[1].end;
strobe[0].end = tstate[5].end;

/* notMemS1 */
strobe[1].start = tstate[1].end;
strobe[1].end = strobe[1].start + strobe[1].data;
if (strobe[1].end > tstate[6].end) strobe[1].end = tstate[6].end;
```

```
/* notMemS2-4, start at least one Tm after start of T2,
   end at end of T5; if configured with zero then never go low */
for (i = 2; i < 5; i++)
 { if (strobe[i].data > 0)
    {
     strobe[i].start = tstate[1].end + strobe[i].data;
     strobe[i].end = tstate[5].end;
    }
  else strobe[i].end = strobe[i].start = 0; /* never low */
 }

/* Read and write strobes */
notMemRd.start = tstate[3].end;
notMemRd.end   = tstate[5].end;

if (early_write) notMemWr.start = tstate[2].end;
         else notMemWr.start = tstate[3].end;
notMemWr.end = tstate[5].end;
}


void print_data()
{ int i;
printf("\n\n   Tstates    \n");
printf("      start      end\n\n");
for (i = 1; i < 7; i++)
   printf("T%d   %d       %d\n",i,tstate[i-1].end, tstate[i].end);

printf("\n\n Strobes \n");
printf("      start     end\n\n");
for (i = 0; i < 5 ; i++)
printf("notMemS%d   %d    %d  \n",i,strobe[i].start,strobe[i].end);
printf("notMemRd   %d    %d\n",notMemRd.start, notMemRd.end);
printf("notMemWr   %d    %d\n",notMemWr.start, notMemWr.end);
}


int main( int argc, char *argv[])
{
printf("Transputer memory configuration program\n");
if (argc <2) exit(0);
read_config(argv[1]);/* read the configuration data file */
set_tstates();       /* set up the end points of the T states*/
set_strobes();       /* set up the end points of the strobes */
print_data();        /* print the results */
return(0);
}
```

Appendix C

Product list

This appendix gives some of the hardware components available at the time of writing (December 1989) and the suppliers involved. Clearly such information will go out of date fairly rapidly, so the list should be considered as a set of guidelines rather than an exhaustive list.

Compute-only TRAMs

INMOS

B401	T414-20, T425-25 or T800-25 32K/3 SRAM, size 1
B402	T222-20, 8K/2 SRAM, size 1
B403	T414-20, T425-20 or T800-20, 1M/3 DRAM, subsystem, size 4
B404	T414, T425 or T800 32K/3 SRAM, 2M/4 DRAM, subsystem, size 2
B405	T800-20, 8M/5 DRAM, subsystem, parity, size 8
B410	T801 160K/2 SRAM, size 2
B411	T425-20 or T800-20, 1M/3 DRAM, size 1
B416	T222, 64K/2 SRAM, size 1
B417	T800, 64K/3 SRAM, 4M/4 DRAM, subsystem, size 4

Transtech

TTM1	T414, T425 or T800 at 20 MHz, 32K/3 SRAM, size 1
TTM2	T414, T425 or T800 at 20 MHz, 128K/3 SRAM, size 1
TTM3	T414, T425 or T800 to 25 MHz, 1M/3 or 1M/4 size 1
TTM4	T414, T425 or T800 at 20 MHz, 1M/4, subsystem, size 4
TTM5	T414, T425 or T800 to 25 MHz, 1M/3 or 1M/4, size 2
TTM6	T414, T425 or T800 to 25 MHz, 2M/3 or 2M/4, subsystem, size 2
TTM7	T414, T425 or T800 to 25 MHz, 1M/3 or 1M/4, subsystem, size 1
TTM8	T414, T425 or T800 to 25 MHz, 4M/3 or 4M/4, subsystem, size 2
TTM9	T414, T425 or T800 to 25 MHz, 4M/3 or 4M/4, subsystem, size 4

TTM15	T414, T425 or T800 to 25 MHz, 4M/3 or 4M/4, subsystem, size 1
TTM16	T414, T425 or T800 to 25 MHz, 4M/3 or 4M/4, size 1
TTM17	T414, T425 or T800 to 25 MHz, 4M/3 or 4M/4, subsystem, upgradable to 8M, size 2
TTM18	T414, T425 or T800 to 25 MHz, 8M/3 or 8M/4, subsystem, size 2

Special application TRAMs

Graphics

INMOS B408: T800-20, 1M/4 DRAM, 1.25M/4 dual-port RAM (use with B409), max. resolution 1024 × 768, size 8.

INMOS B409: T222-20, video timing generator, color lookup tables (use with B408), max. dot rate 64 MHz, size 8.

INMOS B419: T800-20, G300 color video controller, 2M/4 DRAM, 2M/4 VRAM, max. video resolution 1280 × 1024, 8 bits/pixel size 2.

Ethernet

INMOS B407: T222-20, 64K/3 SRAM Am7990 lance.

SCSI

INMOS B422: T222-20, 64K/2 SRAM, subsystem, target/initiator, size 2.

T2 systems: Paradise-1 SCSI size 4, T222.

ROM

INMOS B418: T222-20, 256K flash ROM, subsystem, size 2.

GPIB

INMOS B421: T222-20, 48K SRAM, 8K EEPROM, GPIB controller, size 4.

Link interface

INMOS B415: RS422 buffer for four links, reset and system services, size 1.

Other

INMOS B420: Vector processing TRAM. T800-25, 1M DRAM, 256K dual-port SRAM, vector processor, size 2.

PC boards

Cesius

Scientific accelerator: T800-25, 4 or 16M DRAM.

INMOS

B008-1: Ten TRAM sites, C004 link switch, T212 controller, C012 bus interface.

Transtech

TMB04: T414-20, T425-20, T800-20 or T800-25 plus four TRAM sites. Up to 16 Mbytes RAM, 4 or 3 cycle, C012 bus interface, DMA.

TMB05: M212 disk processor, 20 or 55 Mbytes winchester, four TRAM sites, DMA.

TMB08: Ten TRAM sites, C004 link switch, T222 controller, C012 bus interface, DMA, INMOS B008 compatible.

Microway

Monoputer: T414, 2 Mbytes RAM.

Quadputer: Four T414 or T800, each with 1 or 4 Mbytes RAM.

Videoputer: T800, 1 or 4 Mbytes RAM, 1 Mbyte video RAM, 256 colors from palette of 16.7 million. Max. display resolution 1024 × 1024.

Parsytec

TPM-PC: T800-20 or T414, 256K to 8M DRAM, optional C004.

MTM-PC: Four T800-20 or T414-20, 1M/4 DRAM each, two C004, two TRAM sites.

BBK-PC: Adaptor for adding busless transputer cards to PC (see Parsytec entries under non-bus cards)

Quintek

FAST1S: T425 or T800 1M DRAM.

FAST1XL T425 or T800, 2, 4, 8 or 16 M DRAM.

FAST4: Four T425 or T800, 1M per processor.

FAST9: Nine T425 or T800, 1M per processor, C004.

FAST17: One processor with 4M, sixteen with no external memory.

Harlequin: T800 4 Mbytes RAM and 512K VRAM, display resolution 512 × 512. Frame grabber included.

Sang

MEGA-Link01: One to four T245/T800, 1 or 4 Mbytes per transputer.

MEGA-Link02: T425 or T800, G300, 1–8 Mbytes RAM, 1–2 Mbytes video RAM, 256 colors from 16.7M, max. resolution 1280 × 1024. DMA.

MEGA-Link03: T245/T800 1–32 Mbytes RAM, DMA.

Gemini

GM8101: T800, 4 or 8M DRAM, B004 compatible.

GM8102: T800, 1–16M DRAM, pipeline of five T800 with no memory, C004.

GM8103: T800 2M DRAM, 0.5–4M video RAM, max. video resolution 1536 × 1152, 8 bits/pixel. C012 interface to bus.

GM8104: T800, 1M DRAM, 0.5–4M video display memory, video resolution to 1024 × 768, 32 bits per pixel. Frame grabber.

GM8110: Motherboard. Ten TRAM sites, C004, T212, B008 compatible.

GM8401: DMA link adaptor.

VME and Sun

INMOS

B011-2 6u bus master: T800-20, 2 Mbytes dual-ported RAM, two TRAM slots, two RS232.

B014-1 6u bus slave: Eight TRAM slots, two C004, C012 bus interface.

B016-1 6u bus master/slave: T801-25, 256K static RAM, 4 Mbytes dual-ported RAM, 256 Kbytes PROM, two RS232 ports.

Meiko

In-Sun computing surface boards: One to four T800, up to 4M shared DRAM, 1M dual port to VME bus. Message link switch and system supervisor.

Transtech

MCP501: Eight TRAM slots.

Parsytec

VMTM: Multi-transputer board for VME and Sun, 6u, four T800-20 or T414-20, 1M/4 each, two C004, four link adaptor interface to VME bus.

BBK-V2: Active bus bridge for VME and Sun, 6u, master/slave, T800-20 or T414-20, 2M dual-port DRAM.

BBK-V1: Passive bus bridge for VME and Sun VME DMA controller to C012 link interface.

MTM-SUN: Multi-transputer 9u module for Sun. Four T800-20, 1M each, expandable to 10 with Parsytec cards 64 × 64 link switch, dual-port RAM interface to VME bus.

HP

Protek

TRM12A motherboard for HP series 300 DIO-II backplane, twelve TRAM sites, T212 controller, two C004, link adaptor interface to host.

TRM4A motherboard for HP series 200/300 DIO-I backplane, four TRAM sites, C004, T212 controller, link adaptor interface to host.

PS/2

Parsytec

TPM-PS/2 T800-20 or T414-20, 256K to 4M DRAM, optional C004 daughterboard, three daughterboard sockets.

BBK-PS/2 Active bus bridge for PS/2 microchannel T800-20 or T414-20, 2M dual-port DRAM, DMA.

Quintek

Poppy: One processor, 4M.

Apple Macintosh

Levco

TransLink II: Motherboard, four TRAM sites, C012 interface to bus.

Parsytec

TPM-MAC: Transputer board for MAC T800-20 or T414-20, 256K to 7M DRAM, optional C004 daughterboard, three sockets for daughterboards.

MTM-MAC: Dual transputer board for Apple Mac-II, two T800-20 or T414-20, 2M DRAM per processor. Link adaptor interface to bus.

Digital Equipment Q-bus

Caplin

QT0: Bus interface board, T212 providing four links with separate control signals. 2K dual-port SRAM, 4K EPROM, DMA to Q-bus.

QT1: T414 or T800 128K/3 SRAM, 4M DRAM, subsystem, no bus interface.

QT2: Two T414 or T800, 1 or 2M DRAM each, subsystem, no bus interface.

QT4: Four T414 or T800, 1M each, no bus interface.

QTVIO: Graphics card, T800 2M SRAM, max. resolution 512 × 512 12 bits/pixel, video digitizer, 256 colors from 16.7M, no bus interface.

QTIO: Communications modules with T222, 16K SRAM, 16K EPROM, one parallel, eight serial ports. No bus interface.

QTX: Link configuration module. T222 controlling C004, thirty-two links and eight subsystems. No bus interface.

Parsytec

BBK-MB1: Passive bus bridge to DEC Q-bus DMA from bus to four separate C012 link adaptors with RS-422 link drivers.

NEC-PC

INMOS

B015-1: Five TRAM slots, C012 bus interface.

Acorn Archimedes

Gnome

Transputer baseboard for Archimedes workstation: 1 transputer up to 8 Mbytes TRAM motherboard, four TRAM sites.

Non-bus

INMOS

B012-1: Double extended eurocard, sixteen TRAM slots, two C004, T212 controller.

Transtech

TMB12: Double extended eurocard, sixteen TRAM sites, two C004, T222 controller, INMOS B012 compatible.

Parsytec

Wide range of cards with various different options of processors and memory configurations. Graphics cards and a SCSI interface also available. These cards have a proprietary bus but may be used in a PC via the bus adaptor BBK-PC.

Workstation makers

Atari

ATW standalone workstation with 68000 IO processor, T800, 4M, video with custom blitter, display resolutions from $512 \times 512 \times 32$ bits to $1280 \times 1024 \times 4$ bits. Slots for up to three farm cards of four processors each.

Cogent

Multi-user workstation with two T800 processors per display station and multiple processor base station with Nubus interfaces.

Thema

Thema workstation containing 80286 or 80386 PC board and multiple transputer system board slots which can take various combinations of custom boards, including video, SCSI and compute modules.

Large-scale system makers

Meiko: Computing Surface.

Parsys: SuperNode (SN series).

Parsytec: Supercluster.

Telmat: T.Node.

Company Addresses

Caplin

Caplin Cybernetics Ltd, C26 Poplar Business Park, 10 Prestons Road, London E14 9RL. Tel. 071-538 1716, fax 071 538 4151.

Cesius

Cesius Computer Inc., 2111 Wilson Boulevard, Suite 700, Arlington, VA 22201, USA.

Cesius Ltd, 660 Aztec West, Almondsbury, Bristol BS12 4SD, UK. Tel. Bristol (0454) 612425, fax (0454) 618188.

CSA

Computer System Architects, 950 N. University Avenue, Provo, UT 84704, USA. Tel. (801) 374-2300, fax (801) 374-2306.

Distributed Software Ltd

670 Aztec West, Almondsbury, Bristol BS12 4SD, UK. Tel. Bristol (0454) 612777, fax (0454) 618188.

Gemini

Gemini Computer Systems Ltd, Springfield Road, Chesham, Bucks HP5 1PW, UK. Tel. (0494) 791010, fax (0494) 784545.

Gnome

Gnome Computers Ltd, 16 Histon Road, Cambridge CB4 3LE, UK. Tel. Cambridge (0223) 461520.

Thema

Thema, Röntgenstrasse 31, 7080 Aalen, FRG. Tel. (07361) 44031, fax (07361) 44030.

INMOS

INMOS Ltd, 1000 Aztec West, Almondsbury, Bristol BS12 4SQ, UK. Tel. Bristol (0454) 616616, fax (0454) 617910.

Levco

6160 Lusk Boulevard, Suite C-100, San Diego, CA 92121, USA. Tel. (619) 457-201, fax (619) 457-2325.

Meiko

Meiko Ltd, 650 Aztec West, Bristol BS12 4SD, UK. Tel. Bristol (0454) 616171, fax (0454) 618188.

Meiko Scientific Corp., 400 Oyster Point Boulevard, Suite 523, South San Francisco, CA 94080, USA. Tel. (415) 952-9900, fax (415) 952-7092.

Microway

Microway (Europe) Ltd, 32 High Street, Kingston upon Thames, Surrey KT1 1HL, UK. Tel. 081-541 5466.

MIMD

MIMD Systems Inc., 1301 Shoreway Road, Suite 430, Belmont, CA 94002, USA. Tel. (415) 595-7303, fax (415) 595-8158.

Parsytec

Paracom Inc., Building 9 Unit 60, 245 W. Roosevelt Road, West Chicago, IL 60185, USA. Tel. (312) 293-9500, fax (312) 231-0345.

Paracom GmbH, Jülicher Strasse 338, D-5100 Aachen, FRG. Tel. (241) 166000, fax (241) 166050.

Perihelion

Perihelion Software Ltd, The Maltings, Shepton Mallet, Somerset BA4 5QE, UK. Tel. Shepton Mallet (0749) 344203, fax (0749) 344977.

Protek

10 Grosvenor Place, London SW1X 7HH, UK. Tel. 071-245 6844, fax 071-235 7349.

Quintek

Southfield House, 2 Southfield Road, Westbury-on-Trym, Bristol BS9 3BH, UK. Tel. Bristol (0272) 628196, fax (0272) 628717.

Sang

Sang computersysteme GmbH, Am Wünnesberg 13, 4300 Essen-Haarzzopf, FRG. Tel. (201) 7101191, fax (201) 710410.

Sension

Sension, Denton Drive, Northwich, Cheshire CW9 7LW, UK. Tel. Northwich (0606) 44321.

Strand

Strand Software Technologies, Greycaine Road, Watford, Herts. WD2 4JP, UK. Tel. Watford (0923) 247707, fax (0923) 247836.

Strand Software Technologies Inc., 15220 N W Greenbrier Parkway, Suite 350, Beaverton, OR 97006, US. Tel. (503) 690-9830, fax (503) 690-9797.

T2

T2 systems Ltd, 62 Longmead Avenue, Bishopstoke, Eastleigh, Hampshire SO5 6ET, UK. Tel. Southampton (0703) 641276.

Telmat

Telmat Informatique, ZI rue de l'Industrie, BP 12-68360, Soultz, France. Tel. 89 76 52 20, fax 89 74 27 34.

Transtech

Transtech Devices Ltd, Unit 17, Wye Industrial Estate, London Road, High Wycombe, Bucks. HP11 1LH, UK. Tel. High Wycombe (0494) 464303.

3L

3L Ltd, Peel House, Ladywell, Livingston, West Lothian EH4 6AG, Scotland. Tel. Livingston (0506) 415959, fax (0506) 415944.

References and Bibliography

References

Accetta, M., Baron, R., Golub, D., Rashid, R., Tevanian, A. and Young, Y. (1986) 'Mach: a new kernel foundation for Unix development', in *Proceedings of the USENIX 1986 Summer Conference.*

Chesney, M. and Ganz, R. (1989) 'The computing surface', in *Transputer Applications* (ed. Harp, R. G.), Pitman: London, pp. 142–69.

Flynn, M. J. (1966) 'Very high speed computing systems', *Proc. IEEE*, vol. 54, pp. 1901–9.

Flynn, M. J. (1972) 'Some computer organizations and their effectiveness', *IEEE Trans. Comp.*, vol. C-21, pp. 948–60.

Gelernter, D. (1985) 'Generative communication in Linda', *ACM Trans. Prog. Lang. Sys.*, vol. 7(1), pp. 80–112.

Gelernter, D. (1988) 'Getting the job done', *Byte*, November 1988.

Harp, J. G. (1987) Phase 2 of the reconfigurable transputer project (P1085), *ESPRIT 87 Achievements and Impact*, Part 1, North Holland: Amsterdam.

Harp, J. G., Jesshope, C. R., Muntean, T. and Whitby-Strevens, C. (1987) Phase 1 of the development and application of a low cost high performance multiprocessor machine, *ESPRIT 86 Results and Achievements*, North Holland: Amsterdam.

Hoare, C. A. R. (1985) *Communicating Sequential Processes*, Prentice Hall: Hemel Hempstead.

Hockney, R. W. and Jesshope, C. R. (1988) *Parallel Computers 2*, Adam Hilger: Bristol.

Jones, G. and Goldsmith, M. (1988) *Programming in occam 2*, Prentice Hall: Hemel Hempstead.

Krishnamurthy, E. V. (1989) *Parallel Processing: Principles and practice*, Addison-Wesley, London.

Meiko (1987) *Inter-module Link Interface*, Meiko Technical Note.

Perihelion Software (1989) *The Helios Operating System*, Prentice Hall: Hemel Hempstead.

Reddaway, S. F. (1984), 'Distributed array processor, architecture and performance' in *High-Speed Computing* (ed. Kowalik, J. S.), NATO ASI Series, vol. 7, Springer Verlag: Berlin, pp. 89–98.

Tanenbaum, A. S. (1987) *Operating Systems: Design and implementation*, Prentice Hall: Hemel Hempstead.

Bibliography

Askew, C. (ed.), *Occam and the Transputer: Research and applications. Proceedings of the 9th occam User Group Technical Meeting*, IOS: Amsterdam (1988).

Bakkers, A. (ed.), *Applying Transputer Based Parallel Machines. Proceedings of the 10th occam User Group Technical Meeting*, IOS: Amsterdam (1989).

Harp, G. (ed.), *Transputer Applications*, Pitman: London (1989).

INMOS, *Communicating Process Architecture*, Prentice Hall: Hemel Hempstead (1988).

INMOS, *The occam 2 Reference Manual*, Prentice Hall: Hemel Hempstead (1988).

INMOS, *Transputer Development System*, Prentice Hall: Hemel Hempstead (1988).

INMOS, *The Transputer Instruction Set: A compiler writer's guide*, Prentice Hall: Hemel Hempstead (1988).

INMOS, *Transputer Reference Manual*, Prentice Hall: Hemel Hempstead (1988).

INMOS, *Transputer Technical Notes*, Prentice Hall: Hemel Hempstead (1988).

Kernighan, B. W. and Ritchie, D. M., *The C Programming Language*, Prentice Hall: Englewood Cliffs, NJ (1978).

Kerridge, J. (ed.), *Developments Using occam. Proceedings of the 8th occam User Group Technical Meeting*, IOS: Amsterdam (1988).

Muntean, T. (ed.), *Parallel Programming of Transputer Based Machines. Proceedings of the 7th occam User Group Technical Meeting*, IOS: Amsterdam (1988).

Perrott, R. H., *Parallel Programming*, Addison-Wesley, London (1987).

Wexler, J. (ed.), *Developing Transputer Applications. Proceedings of the 11th occam User Group Technical Meeting*, IOS: Amsterdam (1989).

Index

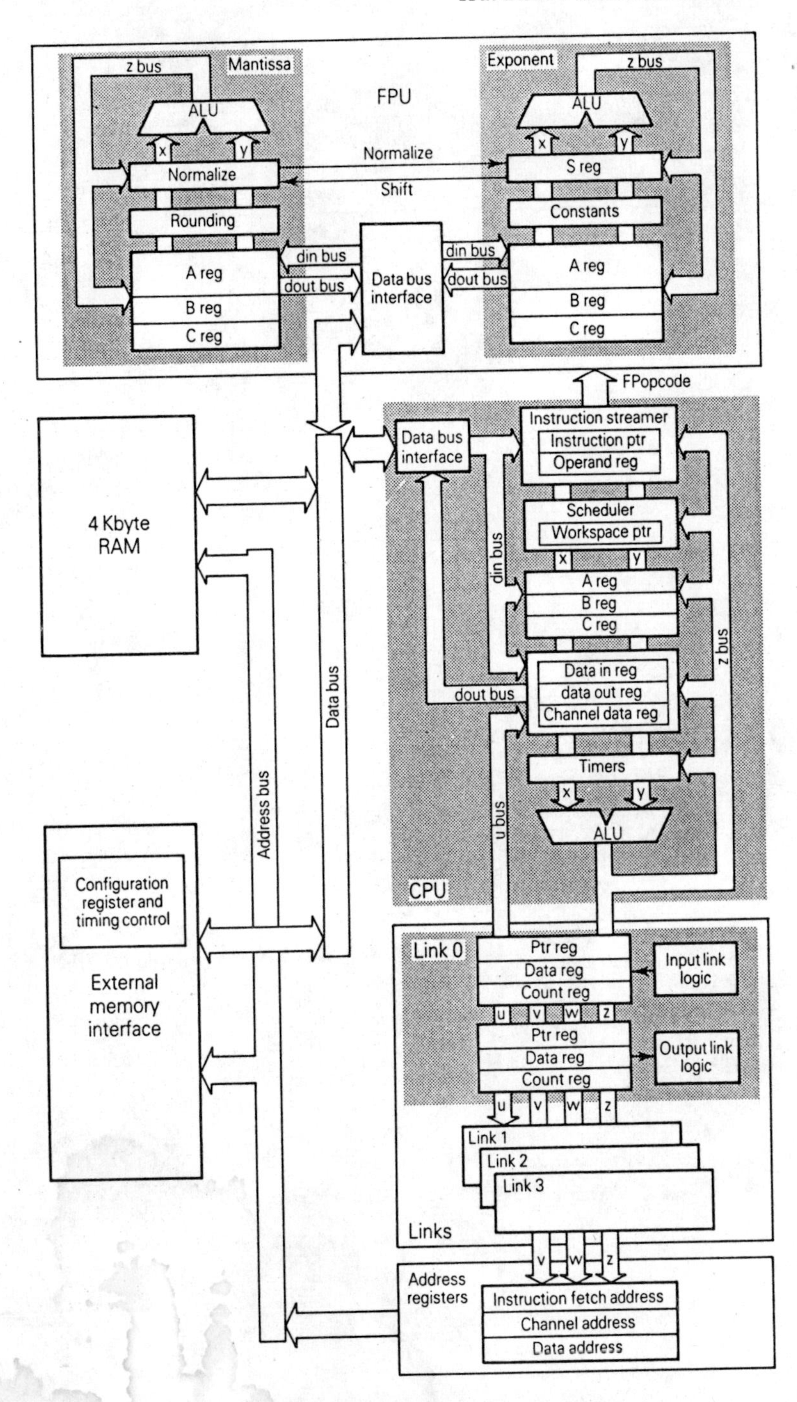

Figure 7.2 Data paths in the T805 processor